D0064784

SCIENCE AND MEDICAL DISCOVERIES

Other books in the
Exploring Science and Medical Discoveries series:

Antibiotics
Gene Therapy
Vaccines

EXPLORING
SCIENCE AND MEDICAL DISCOVERIES

Cloning

Nancy Harris, *Book Editor*

Bruce Glassman, *Vice President*
Bonnie Szumski, *Publisher*
Helen Cothran, *Managing Editor*
David M. Haugen, *Series Editor*

GREENHAVEN PRESS
An imprint of Thomson Gale, a part of The Thomson Corporation

THOMSON
——————✳——————™
GALE

Detroit • New York • San Francisco • San Diego • New Haven, Conn.
Waterville, Maine • London • Munich

11-2-05

THOMSON
GALE

LIBRARY OF CONGRESS CATALOGING-IN-PUBLICATION DATA
Cloning / Nancy Harris, book editor.
p. cm. — (Exploring science and medical discoveries)
Includes bibliographical references and index.
ISBN 0-7377-1965-6 (lib. : alk. paper) — ISBN 0-7377-1966-4 (pbk. : alk. paper)
1. Cloning. I. Harris, Nancy. II. Series
QH442.2.C5643 2005
660.6'5—dc22   2003056829

Printed in the United States of America

CONTENTS

mane treatment animals suffer when being used for cloning research. He claims that animal subjects suffer from tumors, chronic kidney and liver dysfunction, and damaged or missing organs.

Chapter 4: Further Cloning Developments

Most great science and medical discoveries emerge slowly from the work of generations of scientists. In their laboratories, far removed from the public eye, scientists seek cures for human diseases, explore more efficient methods to feed the world's hungry, and develop technologies to improve quality of life. A scientist, trained in the scientific method, may spend his or her entire career doggedly pursuing a goal such as a cure for cancer or the invention of a new drug. In the pursuit of these goals, most scientists are single-minded, rarely thinking about the moral and ethical issues that might arise once their new ideas come into the public view. Indeed, it could be argued that scientific inquiry requires just that type of objectivity.

Moral and ethical assessments of scientific discoveries are quite often made by the unscientific—the public—sometimes for good, sometimes for ill. When a discovery is unveiled to society, intense scrutiny often ensues. The media report on it, politicians debate how it should be regulated, ethicists analyze its impact on society, authors vilify or glorify it, and the public struggles to determine whether the new development is friend or foe. Even without fully understanding the discovery or its potential impact, the public will often demand that further inquiry be stopped. Despite such negative reactions, however, scientists rarely quit their pursuits; they merely find ways around the roadblocks.

Embryonic stem cell research, for example, illustrates this tension between science and public response. Scientists engage in embryonic stem cell research in an effort to treat diseases such as Parkinson's and diabetes that are the result of cellular dysfunction. Embryonic stem cells can be derived from early-stage embryos, or blastocysts, and coaxed to form any kind of human cell or tissue. These can then be used to replace damaged or diseased tissues in those suffering from intractable diseases. Many researchers believe that the use of embryonic stem cells to treat human diseases promises to be one of the most important advancements in medicine.

However, embryonic stem cell experiments are highly controversial in the public sphere. At the center of the tumult is the fact that in order to create embryonic stem cell lines, human embryos must be destroyed. Blastocysts often come from fertilized eggs that are left over from fertility treatments. Critics argue that since blastocysts have the capacity to grow into human beings, they should be granted the full range of rights given to all humans, including the right not to be experimented on. These analysts contend, therefore, that destroying embryos is unethical. This argument received attention in the highest office of the United States. President George W. Bush agreed with the critics, and in August 2001 he announced that scientists using federal funds to conduct embryonic stem cell research would be restricted to using existing cell lines. He argued that limiting research to existing lines would prevent any new blastocysts from being destroyed for research.

Scientists have criticized Bush's decision, saying that restricting research to existing cell lines severely limits the number and types of experiments that can be conducted. Despite this considerable roadblock, however, scientists quickly set to work trying to figure out a way to continue their valuable research. Unsurprisingly, as the regulatory environment in the United States becomes restrictive, advancements occur elsewhere. A good example concerns the latest development in the field. On February 12, 2004, professor Hwang Yoon-Young of Hanyang University in Seoul, South Korea, announced that he was the first to clone a human embryo and then extract embryonic stem cells from it. Hwang's research means that scientists may no longer need to use blastocysts to perform stem cell research. Scientists around the world extol the achievement as a major step in treating human diseases.

The debate surrounding embryonic stem cell research illustrates the moral and ethical pressure that the public brings to bear on the scientific community. However, while nonexperts often criticize scientists for not considering the potential negative impact of their work, ironically the public's reaction against such discoveries can produce harmful results as well. For example, although the outcry against embryonic stem cell research in the United States has resulted in fewer embryos being destroyed, those with Parkinson's, such as actor Michael J. Fox, have argued that prohibiting the development of new stem cell lines ultimately will prevent a timely cure for the disease that is killing Fox and thousands of others.

Greenhaven Press's Exploring Science and Medical Discover-

ies series explores the public uproar that often follows the disclo-
sure of scientific advances in fields such as stem cell research.
Each anthology traces the history of one major scientific or medical discovery, investigates society's reaction to the breakthrough,
and explores potential new applications and avenues of research.
Primary sources provide readers with eyewitness accounts of crucial moments in the discovery process, and secondary sources offer historical perspectives on the scientific achievement and society's reaction to it. Volumes also contain useful research tools,
including an introductory essay providing important context, and
an annotated table of contents enabling students to quickly locate
selections of interest. A thorough index helps readers locate con-
tent easily, a detailed chronology helps students trace the history
of the discovery, and an extensive bibliography guides readers interested in pursuing further research.

Greenhaven Press's Exploring Science and Medical Discoveries series provides readers with inspiring accounts of how generations of scientists made the world's great discoveries possible and
investigates the tremendous impact those innovations have had on
the world.

"With cloning firmly established in the modern canon, it is as if the science and techniques of biology have been liberated from constraints that once seemed inviolable. We and our descendants must wait and see what the world makes of this liberation, or rather, we must try to see that the new power is put only to good and proper use. It would be foolish to underestimate the potential."

—Ian Wilmut and Keith Campbell,
The Second Creation: Dolly and the Age of Biological Control

These words from Ian Wilmut and Keith Campbell, the scientists responsible for the cloning of the famous sheep Dolly, hint at the impact that cloning could have on the world. They also emphasize the caution that most scientists believe should be exercised in regards to cloning technology.

Indeed, precaution may be the wisest way to address society's fears about cloning, especially concerns about human cloning, which suddenly seemed possible following Dolly's birth. However, there is a great divide between the public's fears about how cloning will be used and the actual progress of cloning research and its likely viability. According to most scientists, the technology does not presently exist to safely or reliably perform human reproductive cloning. There are also questions about the viability of so-called therapeutic cloning, cloning for biomedical research, which scientists hope will produce animals to provide such things as transplantable organs for humans. Many scientists say the field does not hold as much potential as previously hoped because of the high expense of cloning procedures and the public's resistance to cloning. Consequently, some scientists are turning to more viable avenues of research, suggesting that both human and therapeutic cloning may be a long way off.

The Science of Cloning

Before exploring society's fears about cloning and the impediments to cloning's success, it is important to understand what cloning is and what the different types of cloning are as well as some of the terminology used. In the most simple and strict sense, the definition of a clone is an organism or group of organisms, genes, or cells that share the same genetic material as a common ancestor. Cloning occurs naturally in nature. For example, people have known since ancient times that simple animals such as starfish and flatworms divide and replicate themselves through asexual reproduction, producing offspring that are exact genetic copies of themselves. Similarly, single-celled organisms such as bacteria and protozoa replicate themselves by fission, or splitting. Some plants have been cloning themselves for billions of years through asexual reproduction or vegetative propagation, but it was not until the early 1900s that the word *clon*, later changed to *clone*, from the Greek word meaning "twig," was used for the first time to designate this process in plants.

Today, in the scientific world, cloning is performed artificially using various techniques. One of these is molecular, gene, or DNA cloning, which refers to the process whereby strings of DNA containing genes are duplicated in a host bacterium. Another is cellular cloning, in which copies of cells are made, resulting in what is called a "cell line," a repeatable procedure where identical copies of the original cell can be grown indefinitely.

Today, when the word *cloning* is used, it usually refers to nuclear transfer, the procedure used to create Dolly. In nuclear transfer, the nucleus of one cell is removed and placed into an unfertilized egg, which has had its nucleus removed. This process, also called somatic cell nuclear transfer (SCNT), is used in reproductive cloning. Reproductive cloning produces a duplicate of an existing animal by placing the clonal embryo in a womb to develop. This process has been used to clone sheep and other animals and is what people fear will be used to clone humans.

SCNT is also used in therapeutic or biomedical cloning. In this procedure, the clonal embryo is used to generate stem cells and is then destroyed. Stem cells, existing in early-stage embryos, are undifferentiated cells that have the potential to form many different types of cells. Scientists believe stem cells may be used to generate treatments for incurable diseases or to develop replacement tissues that would not be rejected by the body's immune system.

Another artificial cloning technique is embryo cloning or artificial twinning. This medical technique duplicates the process nature uses to produce twins. One or more cells are removed from a fertilized embryo and are encouraged to develop into one or more duplicate embryos, which are then placed into a womb. This process creates twins or triplets with identical DNA. A process called embryo twinning can also be done in a laboratory by splitting sexually created embryos into identical halves. This is cloning by fission or cutting of the embryo as opposed to cloning by fusion or nuclear transfer.

Cloning Fears

The latest developments in cloning technology have generated fear and intense debate. Dolly's birth, especially, caused an intense emotional reaction. The sheep's birth made the possibility of human cloning seem more real and stirred reactions worldwide, which were further fueled by sensational news stories. People from all walks of life offered their opinions, expressed their fears, and hotly debated the issue of human cloning. Bioethicists and others, including philosophy professor Barbara MacKinnon of the University of San Francisco, have speculated about what caused the fears. MacKinnon suggests that the deepest fear is the threat cloning seems to pose to the concept of personal identity and uniqueness. She says, "Instead of being a surprising new combination of the genetic contribution of two, the cloned child is somewhat of a repeat of its source. Today we try to see value in diversity and to appreciate differences. Each person is unique." Humans wrestle with their own ideas about what it means to be human and what makes each person special, questions that have puzzled philosophers for centuries. The hypothetical clone appears to be a nonhuman replication and, because it seems to have no unique identity, elicits aversion.

Other fears were raised following Dolly's birth, such as who would or should be cloned and for what purposes. Many worried that scientists would attempt eugenics by altering the genetics of humans through cloning, raising the specter of terrifying armies of Adolf Hitlers or humans created for "spare parts." Others wondered if scientists would upset nature's balance by tampering with the age-old process of human reproduction. There were also concerns that scientists were "playing God," disturbing humanity's subservient relationship to God. The Vatican weighed in on this last fear by condemning human cloning, saying it was leading humankind down a dark path.

Another issue that generated concern was using human embryos in the nuclear transfer process, the first step in both reproductive and therapeutic cloning. In therapeutic cloning the debate focuses on the morality of using the embryo during the cloning procedure and then discarding it. Many believe that the embryo enjoys the same moral status as an adult human and cannot be killed with impunity. Indeed, these critics believe that therapeutic cloning would lead to a devaluation of human life.

Other people worry about the motives of those involved in

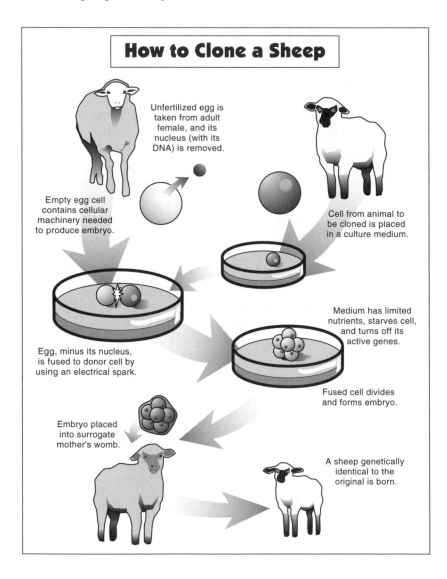

cloning research. To be sure, scientists have used cloning practices for years to propagate farm animals that produce better meat or more milk or animals whose organs will be used for transplantation into humans. Ian Wilmut said he performed the experiments that produced Dolly in order "to make precise genetic changes in cells" so that herds of animals could be created to provide better food or therapeutic proteins to treat human diseases as well as to improve disease resistance in animals. However, motives may sometimes be more financial than humanitarian, as in the case of biotech companies hungering for profits and scrambling for patents on cloning technologies. Indeed, much cloning research is financed with corporate money and conducted by scientists hired by corporations. Their experimental work is conducted in secrecy because of company policies. This situation generates concerns about how the technology will be used, its safety, and how it will be tested and regulated.

Despite these worries, cloning research marches on. Just because something is declared undoable or undesirable does not mean that someone is not going to try it. One such person is fertility expert Dr. Panayiotis Zavos, founder of the Andrology Institute of America, who strongly advocates using human reproductive cloning to assist infertile couples in having children. Zavos thinks that cloning is inevitable, and at the beginning of 2003 he said that he expected to produce a cloned child in the near future. Similarly, several other professionals support reproductive cloning despite the skepticism of experts. As Leon Kass, chairman of President George W. Bush's Council on Bioethics explains:

> The work on the cloning project has proceeded rapidly. "To clone or not to clone a human being" is no longer a fanciful question. Success in cloning sheep, and also cows, mice, pigs, goats, and cats, makes it perfectly clear that a fateful decision is now at hand: whether we should welcome or even tolerate the cloning of human beings. If recent newspaper reports are to be believed, reputable scientists and physicians have announced their intentions to produce the first human clone in the coming year [2003]. Their efforts may already be under way.

Indeed, in January 2003 a biotech company named Clonaid made an announcement claiming it had cloned a human baby girl, creating a new surge of debate in Congress about cloning and prompting some representatives to use the incident to push for a total ban on all cloning. Interestingly, the company never provided evidence of the baby, and most scientists have dismissed the announcement as fraudulent. Still, the announcement generated a new wave of fear, expressed by scientists and nonexperts alike. Microbiologist Norton Zinder expressed his concern about human reproductive cloning, saying that "it would be criminal at this stage in our abilities."

Reaction to Fear: Cloning Regulations and Bans in the United States

Some of these fears following the birth of Dolly had been quickly addressed when President Bill Clinton signed a five-year moratorium on the use of federal funds for human reproductive cloning research. Clinton's U.S. National Bioethics Advisory Commission issued a report in 1997 saying that human reproductive cloning would be unethical and unsafe for a variety of reasons. Harold Shapiro, the commission's chair, said cloning was still "clinically and scientifically premature to produce human infants." Furthermore, he argued, even if the technology was safe, it would be unethical to pursue human cloning without a clearer public consensus. Later, President Bush's Council on Bioethics, after many months of in-depth discussion about the numerous controversies surrounding reproductive and therapeutic cloning, supported a ban on human reproductive cloning as well as a four-year moratorium on therapeutic cloning. In July 2001 the House of Representatives passed a bill that would ban human cloning for reproductive or research purposes. Although President Bush urged the Senate to do the same, as of July 2003, the Senate bill had not passed. The Senate is torn between a complete ban on cloning and alternative measures that would allow the use of cloned human embryos for medical research but prohibit using them for reproductive cloning.

The Real Cloning Picture

Many obstacles stand in the way of scientists realizing their cloning goals. Indeed, there have been complications in cloning every animal species, especially other primates. Because the cloned embryos of primates lack the proteins that enable cells to divide properly, repeated attempts at cloning monkeys, using the procedure that created Dolly, have failed. This would seem to make human cloning a more remote possibility.

Following the birth of Dolly, there was much excitement over what therapeutic cloning might offer, and experimental research exploded as scientists cloned animals such as cows, mice, goats, and pigs. However, scientists reported a high rate of experimental casualties, with only a 2 percent survival rate among cloned animals. In other words, typically in the cloning process, only two or three animals survive out of a hundred. For example, in the cloning experiment that produced Dolly, researchers began with 277 fertilized eggs from which only 29 reached the stage of development necessary to be implanted into the uteri of thirteen sheep. Following implantation, only one sheep, Dolly, was born alive. Because of the procedure's inefficiency, it is costly. The estimated cost of the process was fifty thousand dollars. It is believed that human cloning would be significantly more expensive. Researchers have encountered other problems as well, including pregnancies ending in miscarriages, oversized infants, and animals dying shortly after birth or suffering with developmental difficulties. These problems also add to the cost of the procedure.

These are not the only problems associated with cloning research. Cloning experiments involve very delicate procedures; in any one procedure there is a chance of splicing together DNA or having spontaneous mutations occur that could result in hideous abnormalities. If these mistakes were to remain undetected in the genetic code, a sort of biological time bomb might manifest itself in offspring many years later. There is also a problem with the possible reduction in biodiversity. The human species, like all species, needs to maintain a sufficient level of genetic diversity in order to ensure its ongoing health and survival. Should a sizable portion of the population be cloned from the DNA of only a few individuals, a disease that might normally kill off only a small percentage of a heterogenous population might entirely eliminate the homogenous, cloned population.

For all of these reasons, scientists and the biotech companies supporting them have had less incentive to embrace cloning research as the technology's problems have surfaced. Indeed, it now appears that the initial public outcry against cloning may have been an overreaction in light of the slow progress that has been made in cloning research. Also, fears about cloning's harmful potential seem less urgent now that the government has issued bans and passed regulations controlling cloning technology. However, although the public's fears may have outstripped cloning's potential, it should be remembered that researchers once believed that the cloning procedure that created Dolly was scientifically impossible. Most scientists today are insisting that cloning humans is unsafe and impractical, yet there are those who are proceeding with attempts to clone humans nonetheless. Although a strong argument can be made that humans will never be cloned, Dolly proved that predictions about cloning can sometimes be wrong.

CHAPTER 1

Early Cloning History

Cloning in the World of Plants

By Doris M. Stone

In this selection Doris M. Stone explains that plants have been cloning themselves—in a process called asexual reproduction or vegetative propagation—for billions of years in the natural world. Stone says that clones in the plant world are not identical copies of their parents; when plants propagate, spontaneous changes within the plants' cells occur, creating variations in the new plants. These variations are infinitely small compared to those produced through sexual reproduction in other plants and most animals, Stone explains. Because plant clones have little variability, they also have little adaptability, which means they require a very stable environment. However, there are advantages to vegetative propagation, Stone explains. It is a sure method of reproduction and does not carry the many risks of sexual reproduction. According to Stone, while cloning is a successful way for many plants to reproduce, asexual reproduction would be impossible in the higher animals because their physiologies are too complex. Doris M. Stone is an author and director of education at the Brooklyn Botanical Garden.

Confronted by the wealth of science fiction on the subject, one would think that cloning was a method of reproduction invented by Aldous Huxley[1] and succeeding generations of writers. Animals, it is true, have not gone in much for cloning, at least not in the advanced groups; they have directed all their energies into the sexual method of reproduction, leaving cloning to the primitive forms of animal life. Plants, on the other hand, have

1. Aldous Huxley wrote *Brave New World*, a novel in which people are produced asexually in order to fill specific functions in society.

quietly and unobtrusively been carrying on sophisticated cloning operations for billions of years.

Again, we might think from reading science fiction that all clones must be Xerox copies of their parents. This also is not true. Vegetative propagation, a form of cloning, has produced the navel orange, the pink grapefruit, and the nectarine (whose parent plant is the peach). Such variations from the norm, or *sports*, as they are often called, result from spontaneous changes in the hereditary material within the cell—either within the nucleus itself or in the cytoplasmic genes located within such organelles as chloroplasts and mitochondria. The incidence of such variation is infinitely small compared with that encountered in sexually produced organisms. ... But it does occur, and it is a fact that cannot be overlooked in any serious consideration of cloning or asexual reproduction (the older and currently less fashionable term for the process).

A Wide Range of Propagation

Flowering plants, unlike the highly evolved animal groups, exhibit a whole range of asexual methods of propagation. Curiously enough, this type of reproduction is pretty well confined to herbaceous perennials; trees and annuals rely almost exclusively on the sexual method, as do most shrubs.

Most herbaceous perennials store food in underground parts, which facilitates rapid growth in the spring after winter dormancy. Portions of these underground storage organs tend to become detached (usually by decay of connecting tissues) and, provided each has a growing point or bud, they will give rise to new plants. This form of cloning is called vegetative reproduction. Because of its haphazard nature, bulbous and rhizomatous plants clump together into colonies. Daffodils, grasses, bamboos, and crocuses, among many other plants, exist in these expanding colonies, in which distinctions between individuals become blurred. Since all members of the colony are clones, they resemble one another very closely, except for the very few sports or mavericks that may arise. Age is reckoned in terms of the colony rather than the individual.

Trees and shrubs, on the other hand, do not store food in underground organs but within the tissues of aerial stems—trunks, branches, and twigs. Such parts seldom become new plants without human help, there being little provision for natural vegetative reproduction. Detached twigs (cuttings) will usually sprout roots, if properly cared for, or they may produce new plants when grafted to a nearly related species. But practically always people have to get involved.

Annuals put all their surplus food into seeds. As seeds, they tide over a difficult season—a situation comparable to that of many insects, including crickets, praying mantises, and grasshoppers, which die in the fall leaving only eggs to carry on the species. Vegetative propagation of annuals is impossible, even with human help, at least by conventional methods.

Unlike herbaceous perennials, trees and annuals are distinct individuals from seedling to senescent adult. Their life span is like that of animals, predictable and predetermined, and ranging anywhere from three months to several thousand years, depending upon the species. Like people, individual trees can become famous because of their longevity: the General Sherman Redwood of California and the Middleton Oak of Charleston, South Carolina, come to mind. And, also like people, annuals and trees that have been self-sown have unique genomes. . . .

Other Types of Vegetative Reproduction

The big advantage of vegetative reproduction is that it is a sure method, entirely independent of the vicissitudes that beset reproduction by seeds. One great disadvantage is that all the progeny remain close together, competing for light, mineral salts, and water. The gorgeous clumps of iris, the alpine meadows of crocus, and the yellow drifts of daffodils we admire in springtime illustrate this well.

A few plants reproduce vegetatively without the aid of underground storage organs. Plants with "runners," like the strawberry, carry their offspring some distance from the parent plant. Axillary buds on the parent send out horizontal stems that run along the surface of the soil, producing plantlets at intervals. When the connecting runners decay, the plantlets become independent.

Some succulent species of the *Kalanchoë* and *Bryophyllum* genera produce tiny plantlets at the indentations of the leaf margins. When these are detached they become independent plants, but they cannot of course stray far from the parent plant. Rain may cause them to be propelled a yard or so away.

Gardeners and farmers assist plants in their methods of vegetative propagation by separating the offspring and taking them someplace else, thus avoiding overcrowding. Bulbs, corms [underground stems] and tubers are convenient packages to transport over long distances. During the sixteenth century, when travel was slow and protracted, it was possible to introduce the tulip into Austria from its native Turkey and bring the white potato to Europe from its home in the Andes of South America. Living plants of these species would never have survived the long journeys, but their bulbs and tubers arrived in good health.

Barring freak accidents to the hereditary apparatus of the cell—the chromosomes, chromosomal genes, and cytoplasmic genes—every new individual that arises by these asexual methods of reproduction will be an exact replica of the parent. Provided the environment remains stable and favorable, all is well. But should the environment change, clones, which have virtually no potential for variability (and hence adaptability), are at high risk.

Man has exploited the plant's facility for cloning by reproducing the desirable and unique genotypes that have originated either by chance pollinations or in deliberate breeding experiments. All of the Idaho potatoes, red Delicious apples, Peace roses, loganberries, and common soulange magnolias (*Magnolia X soulangiana*) are clones of original crosses. The cloning is done with stem, root, and leaf cuttings, and through grafting. One advantage breeders of perennials have over animal breeders is that once a new variety or hybrid is produced, it can be perpetuated indefinitely. (Annuals, as already mentioned, cannot be cloned, so breeding new varieties of these is much more complicated; often, as in hybrid corn, the same cross must be repeated every year to obtain the superior seed.) . . .

The Important Diversity Created by Sexual Reproduction

The fact that the animal kingdom does not indulge in cloning leads one to suspect that sexual reproduction must confer some great survival advantage. . . .

The process of sexual reproduction in flowering plants has risks involved: the vicissitudes and uncertainties of pollination and seed dispersal, the vulnerability of seedlings—and the expense in terms of the carbon balance sheet. The cost in food of sexual, as opposed to asexual, reproduction is quoted as on the order of 30 to 1. So why do plants bother? Why become so dependent on animals for help, not to mention the vagaries of the weather? The answer is that the survival of the species is at stake. True seeds of a species, unlike its clones, are highly variable in their characteristics. Not in major characteristics, obviously, but in a whole gamut of minor ones, such as a tendency to be tall or short, hairy or smooth, yellow- or pink-flowered, scented or nonscented, green-leaved or with variegated foliage.

In a changing world this kind of variability has important survival value. Because of this and for other reasons, advanced animal groups have completely abandoned cloning in favor of sexual reproduction. In actual fact, asexual methods of reproduction, such as the budding of the tiny animal hydra, would be impossible in vertebrate animals given the high degree of specialization of their tissues. The anatomy of vertebrates is so complex that no organism can regenerate a lost limb, let alone a complete clone. Plant tissues are neither so highly differentiated nor so specialized, so for them asexual reproduction is a relatively easy process. Still, even with this facility, nearly all groups of plants exhibit some form of sexual reproduction, presumably as insurance against environmental change.

A Short History of Cloning

By Steven L. Baird

In the following selection Steven L. Baird explains that clones are produced both asexually and sexually in the natural world. For example, identical twins are clones as are plant tubers and bulbs. Cloning, Baird explains, is also executed artificially through a process called nuclear transfer. In nuclear transfer, scientists remove the nucleus of an unfertilized egg cell and replace it with the nucleus of a cultured cell, which creates a clone. This process is used in two very different procedures: reproductive cloning, in which a clonal embryo is implanted in a woman's womb with the intent of creating a child, and therapeutic cloning, in which a clonal embryo is used to produce stem cells.

Baird also gives details of the history of artificial cloning, beginning with the German embryologist Hans Spemann who, in 1928, was the first to use nuclear transfer to create a clone from a salamander embryo. Following Spemann, developmental biologists Robert Briggs and Thomas King restimulated scientific interest in cloning by cloning tadpoles from frog embryos using body cells instead of embryonic cells. British biologist John Gurdon, in the early 1960s, continued research on nuclear transfer and produced adult frogs from tadpole intestine cells, proving that even specialized cells could produce a complete organism. Following these experiments, scientists were unsuccessful in cloning vertebrates and turned to the cloning of genes in the 1970s. Baird explains that sheep, cows, pigs, and rabbits were cloned in the 1980s and 1990s, but these animals contained the genetic material of both parents because the embryos were sexually fertilized. It was not until 1996, with the birth of the lamb Dolly, that the first mammalian clone was created from the cell of an adult animal. Scientists at the University of Hawaii quickly followed Dolly's birth using the same process to clone fifty mice.

Steven L. Baird, "Technological Literacy and Human Cloning," *Technology Teacher*, vol. 62, November 2002, p. 19. Copyright © 2002 by the International Technology Education Association. Reproduced by permission.

Steven L. Baird works as a technology education teacher at Bayside Middle School in Virginia Beach, Virginia, and is an adjunct faculty member at Old Dominion University.

Cloning is fundamental to most living things, since the body cells of plants and animals are clones ultimately derived from the mitosis of a single fertilized egg. A clone is the name for a group of organisms or other living matter with exactly the same genetic material. The word "clone" has been applied to cells as well as to organisms, so a group of cells stemming from a single cell is also called a clone. Cloning is the production of an exact genetic duplicate of a living organism or cell. Today, for many people, the word "clone" can be confusing and hard to associate with a simple definition. Cloning occurs often in the natural world. With humans and other higher animals, clones form naturally through genetically identical multiple births. Single-celled organisms, such as bacteria, protozoa, and yeast, produce genetically identical offspring through asexual reproduction. These offspring develop from only one parent and are considered clones. Plants can also reproduce asexually through a process called vegetative propagation. Many plants show this by producing suckers, tubers, or bulbs to colonize the area around the parent. Hydras, flatworms, and other simple animals can be cloned through asexual reproduction or the process of regeneration.

Cloning, as an artificial technique, is achieved through nuclear transfer. Nuclear transfer is taking the nucleus of a cultured cell and transferring it to an unfertilized egg cell, which has had its genetic material removed. To effectively understand the science of human cloning, the term cloning must be further developed to distinguish between two basic applications, reproductive cloning and therapeutic cloning. Reproductive cloning uses the cloning procedure to produce a clonal embryo that is implanted in a woman's womb with intent to create a fully formed living child. Therapeutic cloning uses the cloning procedure to produce a clonal embryo, but instead of being implanted in a womb and brought to term it is used to generate stem cells.

At the heart of these two distinctions between cloning lies the basis for most people's arguments either for or against human cloning. There are those who oppose reproductive cloning but support therapeutic cloning. Others are against both types of cloning because they are either opposed to the destruction of embryos that result from therapeutic cloning or because they feel the acceptance of therapeutic cloning will lead to the acceptance of reproductive cloning and eventually to human genetic manipulation.

A History of Cloning

Since ancient times, people have known that invertebrates (animals without backbones), such as earthworms and starfish, can be cloned simply by dividing them into two pieces. Each piece regenerates into a complete organism. Scientific attempts to clone vertebrates would prove to be much harder. Beginning in the late 1800s, scientists began to question why a cell develops to become specialized in function despite the fact that all cells in an organism originate from the same fertilized egg. In 1902, Hans Spemann, a German embryologist, split a two-celled salamander embryo in two. Following the division, each cell grew to be an adult salamander. Spemann's success with splitting a single cell into two disproved earlier hypotheses that the amount of genetic information carried by a cell diminishes with each division. In 1928, Spemann conducted the first nuclear transfer experiment in which he transferred the nucleus of a salamander embryo cell to a cell without a nucleus. From this single cell grew a normal salamander embryo, proving that the nucleus from an early embryo cell was able to direct the complete growth of a different salamander. Spemann had created a clone. The cloning of higher organisms would be proposed by Spemann as the next logical step; however, he was unable to technically devise a method to attempt any such experiments. No one would succeed in doing so until Robert Briggs and Thomas J. King successfully cloned tadpoles in 1952.

Robert Briggs and Thomas King, development biologists at what is now the Fox Chase Cancer Center in Philadelphia, developed the process of nuclear transfer using body cells from frog embryos to produce tadpoles. Their work stimulated renewed interest within the scientific community, and throughout the 1950s, scientists cloned amphibians such as frogs and salamanders using nuclear transfer. Researchers were unsure of whether the specialization of cells meant that only certain cells had certain genes, or if the genes that were not used by the cell were just inactivated. In the early 1960s, John Gurdon, a British molecular biologist conducting research on nuclear transfer, produced adult frogs from tadpole intestine cells proving that even specialized cells are totipotent (retaining the ability to produce a complete organism). Throughout the 1960s and 1970s, however, efforts by the scientific community to produce a cloned vertebrate that would survive to adulthood would be futile. While creating a viable cloned vertebrate seemed out of reach, cloning would step down to the minute level in 1972, with the first cloning of a gene, (the basic physical and functional units of heredity). The 1970s would also be witness to the injection of human DNA into newly fertilized mouse eggs to produce mice that are part human. When the mice reproduce, they pass their human genetic material to their offspring, creating transgenic mice. Through this procedure, different human diseases can be studied by creating mice with the appropriate genetic composition.

A leap to the 1980s found that the first mammals, sheep, and cows were cloned from embryonic cells. But animals cloned from embryonic cells, through a process called embryo splitting, contain the genetic material of both parents because the embryos are sexually fertilized. These clones from embryonic cells from the same parents fertilized at different times are as different as brothers and sisters. This technique has proved invaluable to livestock breeders and by the 1990s, using this technique, various animals such as pigs, sheep, cows, and rabbits have been cloned.

The world's first mammal cloned from a cell of an adult animal was born in 1996, but her existence wasn't revealed to the world until February 1997. Dolly was cloned from a cell taken from the udder of an adult ewe at the Roslin Institute in Scotland by Ian Wilmut and his colleagues. The following year, scientists at the University of Hawaii cloned more than 50 mice from adult cells, creating three generations of [genetically] identical laboratory animals. At the present time, the biotechnology of cloning is advancing at a feverish pace. Additionally, an article written by Sylvia Pagan Westphal, "'Handmade' Cloning Cheap and Easy," appeared online in NewScientist.com about a newly developed method to create genetically identical copies of animals, cheaper, easier, and better than existing methods.

Until now, the key instrument used in cloning is called the "micromanipulator," an expensive machine that is used by a skilled technician to grab an egg cell under the microscope, insert a very fine needle to suck out its nucleus, and then use another needle to transfer a nucleus from the animal to be cloned. This process is tricky and time consuming, and results are somewhere in the 25% range. In the new technique, egg cells are split in half under a microscope using a very thin blade. The halves are allowed to heal and then a dye is introduced to identify the halves containing the nucleus. The halves containing the nucleus are discarded, leaving only the empty cytoplasts (the cell not containing the nucleus). To create the cloned embryo, a cell from an adult animal is fused first with one cytoplast, then another, by quickly introducing an electric current. This new method of cloning is much cheaper and can be performed without the need of a skilled technician. Another advantage is that this method will be relatively easy to automate, with the end result being mass produced cloned embryos. A major concern of this evolving cloning technique is that its cheapness (the electrofusion machine can be purchased for around $3500) will allow increased attempts at human cloning.

From Science Fiction to Reality

By Lee M. Silver

In this selection Lee M. Silver gives a synopsis of early cloning events and explains how, following the cloning of frog embryos in the early 1960s, cloning became engendered in the popular imagination. Although it did not follow that cloning frog embryos would lead to human cloning, many books and popular films raised the specter of human cloning as an inevitable evil. For example, one movie raised public fears about cloning by depicting scientists cloning an army of Adolf Hitlers. Despite the public attention cloning received at this time, Silver explains that no real scientific progress was made in the field until 1993, when two university scientists cloned human embryos. During the public outcry that followed, the European parliament voted unanimously to ban the procedure and the Vatican called it "a venture into a tunnel of madness." Despite objections such as these, Silver points out that steady technological advancements in cloning technology continued over a fourteen-year period, eventually leading to the cloning of the sheep Dolly from an adult cell in 1996.

Lee M. Silver is a professor in the Departments of Molecular Biology, Ecology, and Evolutionary Biology, and in the program in Neuroscience at Princeton University. Silver conducts research in genetics, evolution, reproduction, developmental biology, and behavioral genetics. He also teaches and lectures widely on the social impact of biotechnology.

The use of "nuclear transplantation" as a means toward the cloning of animals was first developed by Robert Briggs and Thomas King working at the Institute for Cancer Research in Philadelphia during the early 1950s. The frog was chosen for these experiments because its eggs are very large and readily ac-

Lee M. Silver, *Remaking Eden: Cloning and Beyond in a Brave New World*. New York: Avon Books, 1997. Copyright © 1997 by Lee M. Silver. All rights reserved. Reproduced by permission of HarperCollins Publishers.

cessible to manipulation. Although Briggs and King never reached their goal of cloning from adult cells, they set the stage for John Gurdon, who finally succeeded in using this method to obtain tadpoles during the mid-1960s.

The cloning of frogs was never easily accomplished. After transplanting thousands of nuclei extracted from adult skin and gut cells, Gurdon's success rate was still abysmally low, and the few animals he obtained developed only to the tadpole stage before dying. It is certainly possible—and with hindsight, it now seems likely—that Gurdon's difficulties were mainly a consequence of the primitive equipment and technology available at that time. For even a small amount of damage to nuclei or reconstructed eggs could have drastic consequences on development.

But, most scientists interpreted Gurdon's essentially negative results differently. Rather than blaming the technology, we blamed mother nature herself. In an almost religious way, we assumed the existence of a basic biological principle: adult cell nuclei cannot be readily reprogrammed back to an embryonic state. The rare adult donor nucleus that did turn into a tadpole was presumed to have come from an aberrant cell. And if a tadpole could be obtained only rarely, it seemed reasonable to assume that it would never be possible to clone adult cells of more highly developed mammalian species—like human beings—into healthy live-born children.

Indeed, in 1984, when the highly respected embryologist Davor Solter, and his student James McGrath, reported on an extensive series of nuclear transplantation studies—with better equipment and technology—on mouse eggs, their results seemed to validate this basic biological principle. The concluding sentence of their publication in the journal *Science* stated that "the cloning of mammals by simple nuclear transfer is biologically impossible."

Cloning Enters Public Culture

Although scientists viewed Gurdon's results in one light, popularizers of science viewed it in quite another. The fact that even a single frog had been cloned led to the suggestion that cloning *would* be possible with human beings. The idea began to filter into the public consciousness during the late 1960s and was firmly planted there with the 1970 publication of Alvin Toffler's sensational, and still influential, *Future Shock.* Toffler wrote, "One of the more fantastic possibilities is that man will be able to make biological carbon copies of himself. . . . Cloning would make it possible for people to see themselves anew, to fill the world with twins of themselves. . . . There is a certain charm to the idea of Albert Einstein bequeathing copies of himself to posterity. But what of Adolf Hitler?"

Just as the concept of cloning was being absorbed by the public, it was parodied by Woody Allen in his 1973 movie *Sleeper.* Allen plays the mild-mannered Miles Monroe who is transported two hundred years into the future and is mistaken for the chief surgeon charged with the task of bringing back the recently deceased "Leader" of the country. While the Leader has met with an untimely death, his nose has been kept alive for nearly a year through a "massive biochemical effort." Miles Monroe is supposed to clone the Leader's whole body from his nose, as the top biomedical scientists of this future country watch from an operating room observation deck. Allen toys with the dual meaning of life—cellular versus conscious—when his character kidnaps the nose and threatens to shoot it if he is not allowed to go free.

Five years later, the 1978 movie *The Boys from Brazil*, based on a book by Ira Levin, took up Toffler's more menacing idea of a Nazi plot to clone an army of latter-day Adolf Hitlers. And that same year, the J.B. Lippincott Company published a supposed nonfiction book by the science writer David Rorvik entitled *In His Image: The Cloning of a Man.* Rorvik claimed to tell the story of a "worldly, self-educated, aging millionaire" who wanted an heir and succeeded in obtaining "not exactly a son," but rather his genetic equivalent through the use of the same nuclear transplantation technique that John Gurdon had used to clone frogs. Rorvik never provided evidence in support of his claim, and several years later his publisher was forced to admit the book was a hoax.

Cloning Becomes Entrenched in Popular Culture

By the early 1980s, the notion of cloning had become entrenched in popular culture, appearing again and again in movies, television shows, and science fiction novels. And it entered the inanimate world as well, with clones of computers and even perfumes. Clones were seen as almost, but not quite, perfect copies of the original, usually cheaper and assumed to be not as "sharp" in some way.

But even as clones flooded the popular imagination, very little in the way of new scientific results was publicized. Some people knew that frogs had been cloned, but it seemed that no real scientific progress had been made beyond that organism. And then in 1993, the silence was shattered with a report that two George Washington University scientists, Jerry Hall and Robert Stillman, had "cloned human embryos."

The Hall-Stillman experiment caused a brief media stir far out of proportion to what had actually been accomplished. Hall and Stillman had simply taken seventeen early human embryos, between the two-cell and eight-cell stages, removed their zona coats, and then separated each of the cells in each embryo apart from its neighbors. Each individual cell was next surrounded by a synthetic zona coat and allowed to develop in a laboratory dish by itself. After a few days, Hall and Stillman found forty-eight newly formed embryos developing in a normal manner. The experiment was terminated at this point—out of ethical consideration—and the embryos were discarded.

Embryo cloning is a far cry from adult cell cloning. If the Hall-Stillman experiment had been taken to its logical endpoint, it might have been possible to obtain the birth of identical twins or triplets. But even the normal practice of IVF [in vitro fertilization] results in the birth of twins or triplets, albeit nonidentical ones. And the old-fashioned method of reproduction through intercourse produces a million pairs of newborn identical twins, a lesser number of identical triplets, and perhaps a handful of identical quadruplets, around the world each year. So what Hall and Stillman had accomplished in the laboratory was equivalent to a well-known natural process.

Still, even this mimicry of nature provoked immediate outrage from many political corners. The Vatican called it a "perverse choice" and a "venture into a tunnel of madness." Biotech critic Jeremy Rifkin said it heralded "the dawn of the eugenics era," and he organized protest rallies outside the institution where it had taken place. The European Parliament voted unanimously to ban cloning because it was "unethical, morally repugnant, contrary to respect for the person, and a grave violation of fundamental human rights which cannot under any circumstances be justified or accepted." And this was all because two scientists had gently teased single embryos apart into two, three, or four separate cells that grew for a few days by themselves before fading away.

I suspect that if the word *clone* had not been used to describe what Hall and Stillman had done, the media would never have jumped on the story. As it was, two weeks passed between their presentation at a scientific meeting and the first headline: "Scientist Clones Human Embryos and Creates an Ethical Challenge." It was the ominous juxtaposition of those two words—*clones human*—that brought on the hysteria.

From Embryos to Adults

While the cloning of [the sheep] Dolly from an adult cell [in 1996] was unquestionably a giant leap forward in reproductive technology, it was a leap that began from a sturdy platform of technical advances that built quietly upon one another over the preceding fourteen years. The first step was accomplished at the Wistar Institute in Philadelphia in 1983, where Davor Solter and Jim McGrath established a protocol for transferring nuclei from one mouse embryo to another. Their work was critically important for two reasons. First, it demonstrated the general feasiblity of using nuclear transfer technology in mammals. Second, it introduced a modification of the technique used in frogs that greatly increased the rate of embryo survival. Instead of isolating nuclei away from their cellular encasement, as Gurdon had done, Solter and McGrath chose to keep nuclei properly protected within their cytoplasmic environments surrounded by a cellular membrane.

The actual protocol began with the removal and elimination of the nuclei that were already present within the recipient embryo. Then the donor cell was placed in the space between the zona coat and the embryo itself, and the two cells were induced to fuse with a special chemical agent or an electrical pulse.

Although they referred to this protocol as "nuclear transplantation"—and it has been referred to in this manner ever since— Solter and McGrath never transplanted nuclei directly into recipient embryos. Rather, they implanted donor cells next to embryos and then allowed a fusion event to bring the donor nucleus into the cytoplasm of the recipient cell. By keeping donor cells intact until the moment of fusion, Solter and McGrath succeeded in protecting the genetic material within. Their protocol was so efficient and safe that 90 percent of embryos reconstructed with nuclei from other early embryos survived and developed properly.

The next advance on the way to Dolly was accomplished in 1986 by Steen Willadsen, who was working at the ARFC Institute of Animal Physiology in Cambridge, England. What Willadsen did differently from Solter and McGrath was to use nuclear-free *unfertilized* eggs, rather than one-cell embryos, as recipients for donor nuclei. The logic behind this decision is based on the notion that an unfertilized egg is chock full of signal proteins waiting patiently to pounce onto the naked DNA that it expects to receive from the fertilizing sperm cell. And if the egg is presented with a donor nucleus instead, the egg's signal proteins won't know the difference—they'll blindly *try* to pounce onto the donor cell DNA with the same vengeance. This logic was validated when Willadsen reported the birth of healthy lambs that had been cloned from donor cells derived from 8-cell embryos.

Eight more years went by before another important advance in cloning was made by Neal First at the University of Wisconsin in 1994. This time the species was the cow, the donor cells were obtained from an even later embryonic stage, and four calves were born. What First didn't realize, however, was the probable reason for his success. It turns out that a technician in First's laboratory had mistakenly not provided the donor embryo cells with nourishing serum that all cells need to grow properly. As a result, the donor cells stepped out of their normal cycle of growth and division and paused in a type of hibernation phase known to scientists as G0. Could it be that cells in this special state of hibernation might be more amenable to cloning than other cells? Perhaps the signal proteins sitting on the DNA in these cells are more easily dislodged by the ones waiting in the egg cytoplasm.

Using Advanced Donor Cells

Keith Campbell and Ian Wilmut at the Roslin Institute in Edinburgh, Scotland, were intrigued by this possibility, and they set about trying to test it with their favorite animal, the sheep. They easily obtained lambs after nuclear transplantation from nine-day-old embryo donor cells, and they extended their success to donor cells obtained from embryo-like cultures grown over a period of weeks in a laboratory dish. They reported their results in a March 1996 paper entitled "Sheep Cloned by Nuclear Transfer from a Cultured Cell Line." And then they moved on to more advanced donor cells, using precisely the same techniques.

Dolly was born at 5:00 P.M. in the afternoon on July 5, 1996. She resulted from the fusion of a nuclear-free unfertilized egg with a donor cell obtained from the mammary gland of a six-year-old ewe. She was the first mammal to be cloned from an adult cell, and is a generation removed from the fertilization event that actually brought together the gametes from her genetic parents.

Dolly's existence was announced to the scientific community in a paper published in the journal *Nature* on February 27, 1997. Unnoticed in the commotion surrounding this one lamb is the fact that two others were also cloned from skinlike cells obtained from a fetus. The birth and survival of three healthy lambs from highly differentiated donor cells provides a clear demonstration that the cloning of a lamb was not a fluke.

CHAPTER 2

Cloning Dolly

The Creation of Dolly

By Michael Specter and Gina Kolata

According to Michael Specter and Gina Kolata in the following selection, the sensational news of the first mammal cloned from an adult was announced to the world on February 22, 1997. The lamb, Dolly, had been created from the udder cells of a six-year-old ewe in January 1996. Scientists working quietly for years had achieved a feat believed to be impossible by most in the scientific community, Specter and Kolata explain. The news touched off the centuries-old fear surrounding the possibility of human cloning. Ironically, lead scientist Ian Wilmut's purpose was not to make carbon copies of animals but rather to discover how to alter their genetic makeup in order to produce livestock capable of producing proteins humans could consume to improve health.

Specter and Kolata report that the road to Dolly was a long one, beginning in 1938 with the nuclear transfer experiments of German embryologist Hans Spemann and continuing with the more sophisticated cloning of frogs in the 1970s by John Gurdon. Although scientists believed that these experiments could not successfully be duplicated using mammals, a few researchers forged on, including biologist Keith Campbell. Campbell convinced his colleagues at the Roslin Institute in Scotland to try experiments that eventually led to the creation of Dolly. Unfortunately, Dolly was the only lamb to survive from 277 eggs that had been used in the experiment, leading to questions about the viability of cloning mammals.

Michael Specter is a staff writer for the *New Yorker* magazine and Moscow bureau chief for the *New York Times*. A former writer for *Science* magazine, Gina Kolata is a science reporter for the *New York Times*. She is also the author of *Clone: The Road to Dolly and the Path Ahead.*

harles Darwin was so terrified when he discovered that mankind had not been specially separated from all other animals by God that it took him two decades to find the courage to publish the work that forever altered the way humans look at life on Earth. Albert Einstein, so outwardly serene, once said that after the theory of relativity stormed into his mind as a young man, it never again left him, not even for a minute.

But Dr. Ian Wilmut, the 52-year-old embryologist who astonished the world on February 22 [1997] by announcing that he had created the first animal cloned from an adult—a lamb named Dolly—seems almost oblivious to the profound and disquieting implications of his work. Perhaps no achievement in modern biology promises to solve more problems than the possibility of regular, successful genetic manipulation. But certainly none carries a more ominous burden of fear and misunderstanding.

"I am not a fool," Dr. Wilmut said . . . in his cluttered lab, during a long conversation in which he reviewed the fitful 25-year odyssey that led to his electrifying accomplishment and unwanted fame. "I know what is bothering people about all this. I understand why the world is suddenly at my door. But this is my work. It has always been my work, and it doesn't have anything to do with creating copies of human beings. I am not haunted by what I do, if that is what you want to know. I sleep very well at night."

Yet by scraping a few cells from the udder of a 6-year-old ewe, then fusing them into a specially altered egg cell from another sheep, Dr. Wilmut and his colleagues at the Roslin Institute . . . , seven miles from Edinburgh [Scotland] have suddenly pried open one of the most forbidden—and tantalizing—doors of modern life.

People have been obsessed with the possibility of building humans for centuries, even before Mary Shelley wrote "Frankenstein" in 1818. Still, so few legitimate researchers actually thought it was possible to create an identical genetic copy of an adult animal that Dr. Wilmut may well have been the only man trying to do it, a contrast with the fiery competition that has become the hallmark of modern molecular biology.

Dr. Wilmut, a meek and affable researcher who lives in a village where sheep outnumber people, grew more disheveled and harried as the pressure-filled week wore on. A $60,000-a-year government employee at the institute, Scotland's leading animal research laboratory, Dr. Wilmut does not stand to earn more than $25,000 in royalties if his breakthrough is commercially successful.

"I give everything away," he said. "I want to understand things."

Dr. Wilmut has made no conscious effort to improve on science fiction in his work; he said, in fact, that he rarely read it. A quiet man whose wife is an elder in the Church of Scotland but who says he "does not have a belief in God," Dr. Wilmut is the least sensational of scientists. Asked the inevitable questions about cloning human beings, he patiently conceded that it might now become possible but added that he would "find it repugnant."

Dr. Wilmut's objectives have always been prosaic and direct: he has spent his life trying to make livestock healthy, more efficient and better able to serve humanity. In creating Dolly, his goal—like that of many other researchers around the world—was to turn animals into factories churning out proteins that can be used as drugs. Even though the work is early and tentative, and it needs many improvements before it can be used, no scientists have stepped forward to say that they doubt its authenticity.

Many scientists say they are certain that the day will eventually come when humans can also be cloned. Already, scientists in Oregon say they have cloned rhesus monkeys from very early embryo cells. That is not the same as cloning the more sophisticated cells of an adult animal, or even a developing fetus. But any kind of cloning in primates brings the work closer to human beings. That is why what has happened [in Scotland] has rapidly begun to resonate far beyond the tufted glens and heather hills of Scotland. In much the way that the Wright Brothers at Kitty Hawk freed humanity of a restriction once considered eternal, human existence suddenly seems to have taken on a dramatic new dimension.

The eventual impact of this particular experiment on business and science may not be known for years. But it will almost certainly cast important new light on basic biological science.

Already, even the simplest questions about the creation of Dolly provoke answers that demonstrate how profound and novel the research here has been. Asked if the lamb should be considered 7 months old, which is how long she has been alive, or 6 years old, since it is a genetic replica of a 6-year-old sheep, Dr. Wilmut's clear blue eyes clouded for a moment. "I can't answer that," he said. "We just don't know. There are many things here we will have to find out."

The Scots have an old tongue twister of an adage that says "many a mickle make a muckle," or, little things add up to big things. It is certainty true of the cloning of Dolly, who had her conceptual birth in a conversation in an Irish bar more than a decade ago and who was born after a series of painstaking experiments, years of doubt and several final all-night vigils—one bleating little lamb among nearly 300 abject failures.

While the world has become transfixed by the idea of creating identical copies from frozen cells, that was not the result that Dr. Wilmut, or any other scientists interviewed for this article, considers the most significant part of the research.

The true object of those years of labor was to find better ways to alter the genetic makeup of farm animals to create herds capable of providing better food or any chemical a consumer might want. In theory, genes could be altered so animals would produce better meat, eggs, wool or milk. Animals could be made more resistant to disease. Researchers here even talk about breeding cows that could deliver low-fat milk straight from the udder.

"The overall aim is actually not, primarily, to make copies," Dr. Wilmut said, interrupted constantly by the institute's feed mill as it noisily blew off steam. "It's to make precise genetic changes in cells."

Obscure as he may seem to those outside his field, Ian Wilmut has been quietly pushing the borders of reproductive science for decades. In 1973, having just completed his doctorate at Cambridge, he produced the first calf born from a frozen embryo. Cows give birth to no more than 5 or 10 calves in a lifetime. By taking frozen embryos produced by cows that provide the best meat and milk, thawing them and transferring them to surrogate mothers, Dr. Wilmut enabled cattle breeders to increase the quality of their herds immensely.

Since then, while always harboring at least some doubt that cloning was really possible, he has struggled to isolate and transfer genetic traits that would improve the utility of farm animals.

A Spark That Lit

In 1986, while in Ireland for a scientific meeting, Dr. Wilmut heard something during a casual conversation in a bar that caught his attention and convinced him that cloning large farm animals was indeed possible. "It was just a bar-time story," he recalled . . . , in the slight brogue he has acquired after living here for 25 years. "Not even straight from the horse's mouth."

What he heard was the rumor—true, it turned out—that another scientist had created a lamb clone from an already developing embryo. It was enough to push him in a direction that had already been abandoned by most of his colleagues.

By the early 1980's, many researchers had grown discouraged about the practicalities of cloning because of a hurdle that had come to seem insurmountable. Every cell in the body originates from a single fertilized egg, which contains in its DNA all the information needed to construct a whole organism. That fertilized egg cell grows and divides. The new cells slowly take on special properties, developing into skin, or blood or bones, for example. But each cell, however specialized, still carries in its nucleus a full complement of DNA, a complete blueprint for an organism.

The problem for scientists was stark and unavoidable: It was assumed that the nucleus of a mature cell, which has developed, or differentiated, so it could carry out a specific function in the body, simply could not be made to function like the nucleus of an embryo that had yet to begin the process of learning to play its special role. Even though the DNA, with all the necessary genes, was in the differentiated cell, the issue was how to turn it on so it would direct the process of growth that begins with the egg. The essential question for cloning researchers was whether the genes in an adult cell could still be used to create a new animal with the same genes.

The pivotal rumor Dr. Wilmut heard at the meeting in Ireland was that a Danish embryologist, Dr. Steen M. Willadsen, then working at Grenada Genetics in Texas, had managed to clone a sheep using a cell from an embryo that was already developing.

The story, which came from a veterinarian named Geoff Mahon, who worked at the same company, went beyond the research that Dr. Willadsen would publish later that year on cloning sheep from early embryos. Dr. Willadsen said in a telephone interview from his home in Florida that he had indeed done the more advanced work but had never published it.

What he did publish was the result of successfully cloning sheep from very early embryo cells: the first cloning of a mammal. Dr. Willadsen tried that experiment with three sheep eggs. In each case, he removed the egg's nucleus, with all its genetic information, and fused that egg, now bereft of instructions on how to grow, with a cell from a growing embryo. If the egg could use the other cell's genetic information to grow itself into a lamb, the experiment would be a success. It worked.

"The reality is that the very first experiment I did, which involved only three eggs, was successful," Dr. Willadsen said. "It gave me two lambs. They were dead on arrival, but the next one we got was alive." The paper was eventually published in *Nature*, the influential British science journal that . . . published Dr. Wilmut's news of Dolly, and it created a sensation.

But it was the rumor of the unpublished work that captivated Dr. Wilmut. If it was possible to clone using an already differentiated embryonic cell, it was time to take another look at cloning an adult, Dr. Wilmut decided. "I thought if that story was true— and remember, it was just a bar-time story—if it was true, we could get those cells from farm animals," he said. And, he thought, he might even be able to make copies of animals from more mature embryos or eventually from an adult.

When Dr. Wilmut flew back to Scotland, he was already dreaming of Dolly. When he was flying back over the Irish Sea with a colleague, he said, "we were already making plans to try to get funds to start this work."

From Daydreams to Successes

Dr. Wilmut's dominance of the field grew from that day, almost by default. He was nearly alone, out on a limb. His tumultuous field seemed to have run out of steam. Many of its leaders and its students had departed, going to medical school and becoming doctors or accepting lucrative positions at in vitro fertilization centers, helping infertile couples have babies. Two of its stars published a famous paper concluding that cloning an adult animal was impossible, dashing cold water on their eager colleagues. Companies, formed in a flush of enthusiasm a decade earlier, folded by the early 1990's.

Most of the few cloning researchers left were focused on a much easier task. They were cloning cells from early embryos that had not yet specialized. And even though some had achieved stunning successes, none were about to try cloning an adult or even cells from mature embryos. It just did not seem possible.

The idea of cloning had tantalized scientists since 1938. When no one even knew what genetic material consisted of, the first modern embryologist, Dr. Hans Spemann of Germany, proposed what he called a "fantastical experiment": taking the nucleus out of an egg cell and replacing it with a nucleus from another cell. In short, he suggested that scientists try to clone.

But no one could do it, said Dr. Randall S. Prather, a cloning researcher at the University of Missouri in Columbia, because the technology was not advanced enough. It would be another 14 years before anyone could try to clone, and then they did it with frogs, whose eggs are enormous compared with those of mammals, making them far easier to manipulate. Dr. Spemann, who died in 1941, never saw his idea carried to fruition.

In fact, frogs were not successfully cloned until the 1970's. The work was done by Dr. John Gurdon, who now teaches at Cambridge University. Even though the frogs never reached adulthood, the technique used was a milestone. He replaced the nucleus of a frog egg, one large cell, with that of another cell from another frog.

It was the beginning of nuclear transfer experiments, which had the goal of getting the newly transplanted genes to direct the development of the embryo. But the frog studies seemed to indicate that cloning could go only so far. Although scientists could transfer nuclei from adult cells to egg cells, the frogs only developed to tadpoles, and they always died.

Most researchers at the time thought even that sort of limited cloning success depended on something special about frogs. "For years, it was thought that you could never do that in mammals," said Dr. Neal First of the University of Wisconsin, who has been Dr. Wilmut's most devoted competitor.

A Crushing Blow

In 1981, after some rapid advances in technology, two investigators published a paper that galvanized the world. It seemed to say that mammals could be cloned—at least from embryo cells. But, in a crushing blow to those in the field, the research turned to be a fraud.

The investigators, Dr. Karl Illmensee of the University of Geneva and Dr. Peter Hoppe of the Jackson Laboratory in Bar Harbor, Me., claimed that they had transplanted the nuclei of mouse embryo cells into mouse eggs and produced three live mice that were clones of the embryos. Their mice were on the cover of the prestigious journal *Science*, and their work created a sensation.

"Everyone thought that article was right," said Dr. Brigid Hogan, a mouse embryologist at Vanderbilt University in Nashville. Dr. Illmensee, the senior author, "was getting enormous publicity and exposure, and accolades," Dr. Hogan said.

Two years later, however, two other scientists, Dr. James Mc-Grath and Dr. Davor Solter, working at the Wistar Institute in Philadelphia, reported in *Science* that they could not repeat the mouse experiment. They concluded their paper with the disheart-ening statement that the "cloning of mammals by simple nuclear transfer is impossible." After a lengthy inquiry, it was discovered that Dr. Illmensee had faked his results.

Leaders in the field were shattered. Dr. McGrath gave up cloning, got an M.D. degree and is now a genetics professor at Yale University. Dr. Solter gave up cloning and is now the direc-tor of the Max Planck Institute in Freiburg, Germany. Most re-search centers abandoned the work completely.

"Man, it was depressing," said Dr. James M. Robl, a cloning re-searcher at the University of Massachusetts in Amherst. After the paper by Dr. Illmensee, "we all thought we would be cloning an-imals like crazy." Dr. Robl said. He had pursued research to try to clone cows and pigs. Suddenly, it seemed as though he was wast-ing his time.

"We had a famous scientist come through the lab," Dr. Robl said. "I showed him with all enthusiasm all the work I was doing. He looked at me with a very serious look on his face and said, 'Why are you doing this?'"

Forging On

But not everyone was despondent. A few investigators forged on. One of them was Dr. Keith Campbell, a charismatic 42-year-old biologist at the institute here [in Scotland] who specializes in studying the life cycle of the cell. Dr. Campbell, who joined the institute in 1991, said in an interview . . . , "I always believed that if you could do this in a frog, you could do it in mammals." Dr. Campbell, who said he had enjoyed the cloning fantasy "The Boys From Brazil," responded to questions about his earlier work on cloning in an interview [in the summer of 1996], saying "We're only accelerating what breeders have been doing for years."

Soon he had convinced his colleagues at the institute to try the experiments that eventually led to their success with Dolly. "But at that point, we still had much to learn," he said.

The most important step would be to find a way to grow clones from cells that had already developed beyond the very earliest embryonic stage. Whenever cloning had been tried with more specialized cells in the past, it had ended in failure. Until Dolly was born, nobody could be sure whether those failures were because older cells have switched off some of their genes for good or because nobody knew how to make them work properly in an egg.

Because no one knew whether cloning was even possible, it was hard to speculate about what the hurdles might be. But Dr. Campbell had what turned out to be the crucial insight. It could be, he realized, that an egg will not take up and use the genetic material from an adult cell because the cell cycles of the egg and the adult cell might be out of synchrony. All cells go through cycles in which they grow and divide, making a whole new set of chromosomes each time. In cloning, Dr. Campbell speculated, the problem might be that the egg was in one stage of its cycle while the adult cell was in another.

Dr. Campbell decided that rather than try to catch a cell at just the right moment, perhaps he could just slow down cellular activity, nearly stopping it. Then the cell might rest in just the state he wanted so it could join with an egg.

"It dawned on me that this could be a beneficial way of utilizing the cell cycle," he said, in what may turn out to be one of scientific history's great understatements.

What he decided to do was to force the donor cells into a sort of hibernating state, by starving them of some nutrients.

In Wisconsin, Dr. First had actually beaten the Scottish group to cloning a mammal from cells from an early embryo; that occurred when a staff member in the laboratory forgot to provide the nourishing serum, inadvertently starving the cells. The result, in 1994, was four calves. But even Dr. First and his colleagues did not realize the significance of how the animals had been created.

Two years later, Drs. Wilmut and Campbell tried the starvation technique on embryo cells to produce Megan and Morag, the world's first cloned sheep and, until now, the most famous sheep in history. Their creation really laid the foundation for what happened with Dolly, for Dr. Campbell succeeded in doing an end run around the problem of coordinating the cycles of the donor cell with the recipient egg.

Today, Megan and Morag munch contentedly in the same straw-covered pen with the new star of the Roslin Institute, angelic little Dolly. Megan and Morag seem completely normal, if slightly spoiled.

Megan is now expecting, and she got pregnant the old-fashioned way. "It will always be the preferred way of having children," Dr. Campbell said jokingly. "Why would anyone want to clone, anyway? It's far too expensive and a lot less fun than the original method."

When the scientists moved on to cloning a fully grown sheep, they decided to use udder, or mammary, cells, and that is how Dolly got her name. She was named after the country singer Dolly Parton, whose mammary cells, Dr. Wilmut said, are equally famous.

How the Experiment Worked

In the experiment that produced Dolly, Dr. Wilmut's team removed cells from the udder of a 6-year-old sheep. The cells were then preserved in test tubes so the investigators would have genetic material to use in DNA fingerprinting—required to prove that Dolly was indeed a clone. In fact, by the time Dolly was born, her progenitor had died.

The trick, Dr. Wilmut said, was the starvation of the adult cells. "You greatly reduce serum concentration for five days," Dr. Wilmut said, "That's the novel approach. That's what we submitted a patent for." And that is why the team was silent about the lamb's birth for months. Until the patent was applied for, nobody wanted the news to spread.

But success is a relative concept. Even Dr. Campbell's technique has failed far more often than it has succeeded. Dolly was the only lamb to survive from 277 eggs that had been fused with adult cells. Nobody knows, or can know, until the work is repeated, whether the researchers were lucky to get one lamb—whether in fact that one lamb was one in a million and not just one in 277 or whether the scientists will become more proficient with more refinement.

The cell fusion that produced Dolly was done in the last week of January 1996. When the resulting embryo reached the six-day stage, it was implanted in a ewe. Dolly's existence as a growing fetus was first discovered on March 20, the 48th day of her surrogate mother's pregnancy. After that, the ewe was scanned with ultrasound, first each month and then, as interest grew, every two weeks.

"Every time you scanned, you were always hoping you were going to get a heartbeat and a live fetus," said John Bracken, the researcher who monitored the pregnancy.

"You could see the head structure, the movement of the legs, the ribs," he said. "And when you actually identified a heart that was beating, there was a great sense of relief and satisfaction. It was as normal a pregnancy as you could have."

On July 5 at 4 P.M., Dolly was born in a shed down the road from the institute. Mr. Bracken, a few members of the farm staff and the local veterinarian attended. It was a normal birth, head and forelegs first. She weighed 6.6 kilograms, about 14½ pounds, and she was healthy.

Because it was summer, the few staff members present were very busy. There was no celebration.

"We phoned up the road to inform Ian Wilmut and Dr. Campbell," Mr. Bracken said.

But Dr. Wilmut does not remember the call. He does not even remember when he heard about Dolly's birth.

"I even asked my wife if she could recall me coming home doing cartwheels down the corridor, and she could not," he said.

Dolly's Significance: Propelling Humanity into a New Age

By Ian Wilmut, Keith Campbell, and Colin Tudge

Ian Wilmut, the lead scientist responsible for the cloning of the famous sheep Dolly in 1996, explains in the following selection, which was excerpted from *The Second Creation*, a book he wrote with Keith Campbell and Colin Tudge, that cloning Dolly overturned one of the deepest dogmas held in biology. Up until Dolly's creation, scientists believed it was "biologically impossible" to produce a mammal from the adult body cell of another mammal. Wilmut and his colleagues dispelled this long-held belief and ushered in what Wilmut calls "the age of biological control." Wilmut points out, however, that cloning is still surrounded by misconceptions; for example, many people believe that exact replications of mammals, including humans, is possible. One reason exact replication is not possible, he explains, is that genes alone do not determine every detail of an animal's (or human's) physique and personality. Therefore, transference of genetic material from one animal to another does not mean that an exact duplicate of the entire organism will be produced. Wilmut further states that human cloning was never the goal of his team; rather, Wilmut's interest lay in genetically engineering animals for use in agriculture, medicine, and conservation.

Ian Wilmut is a Scottish embryologist, professor, and joint head of the Department of Gene Expression and Development at the

51

Roslin Institute near Edinburgh, Scotland. He was the leader of the team that cloned Dolly. Keith Campbell is an English cell biologist and embryologist working as a professor of animal development at the University of Nottingham where he also does research. Campbell joined the Roslin Institute in 1990 to work with Wilmut on the project that resulted in Dolly. Colin Tudge is a science journalist and broadcaster as well as a research fellow at the Center for Philosophy at the London School of Economics. He is the author of more than a dozen books, including *The Variety of Life: A Survey and Celebration of All the Creatures That Ever Lived.*

Dolly seems a very ordinary sheep—just an amiable Finn-Dorset ewe—yet as all the world has acknowledged, if not entirely for the right reasons, she might reasonably claim to be the most extraordinary creature ever to be born. Mammals are normally produced by the sexual route: an egg joins with a sperm to form a new embryo. But in 1996 Keith Campbell and I, with our colleagues at Roslin Institute and PPL [Pharmaceutical Proteins Ltd.], cloned Dolly from a cell that had been taken from the mammary gland of an old ewe and then grown in culture. The ewe, as it happened, was long since dead. We fused that cultured cell with an egg from yet another ewe to "reconstruct" an embryo that we transferred into the womb of a surrogate mother, where it developed to become a lamb. This was the lamb we called Dolly: not quite the first mammal ever to be cloned, but certainly the first to be cloned from an adult body cell. Her birth overturns one of the deepest dogmas in all of biology, for until the moment in February 1997 when we made her existence known through a brief letter in the scientific journal *Nature*, most scientists simply did not believe that cloning in such a way, and from such a cell, was possible. Even afterward, some doubted that we had done what we claimed.

Dolly's impact was extraordinary. We expected a heavy response—the birth of Megan and Morag in 1995 had provided some warning of what might follow—but nothing could have prepared us for the thousands of telephone calls (literally), the scores of interviews, the offers of tours and contracts, and in some cases the opprobrium, though much less of that than we might have feared. Everyone, worldwide, knew that Dolly was important. Even if they did not grasp her full significance (and the full significance, while not obvious, is far more profound than is generally appreciated), people felt that life would never be quite the same again. And in this they are quite right.

The Issue of Human Cloning

Most obviously—and unfortunately, because it is certainly not the most important aspect—commentators the world over immediately perceived that if a sheep can be cloned from a body cell then so can people. Many hated the idea, including President [Bill] Clinton of the United States, who called for a worldwide moratorium on all cloning research. But others welcomed human cloning, and some—like Dr. Richard Seed—who is in fact a physicist, not a physician—even offered to set up cloning clinics, surely jumping the gun by several decades since very few scientists have the necessary expertise, and even in the best hands, human cloning at this stage would be absurdly risky. I fielded many of the telephone calls that flooded into Roslin Institute in the days after we went public with Dolly, and quickly came to dread the pleas from bereaved families, asking if we could clone their lost loved ones. I have two daughters and a son of my own and know that every parent's nightmare is to lose a child, and what parents would give to have them back, but I had and have no power to help. I suppose this was my first, sharp intimation of the effect that Dolly could have on people's lives and perceptions. Such pleas are based on a misconception: that cloning of the kind that produced Dolly confers an instant, exact replication—a virtual resurrection. This simply is not the case. But the idea is pervasive and was reflected in articles and cartoons around the world. *Der Spiegel*'s cover showed a regiment of Einsteins, Claudia Schiffers, and Hitlers—the clever, the beautiful, and the not very nice.

Yet human cloning is very far from Keith's and my own thoughts and ambitions, and we would rather that no one ever attempted it. If it is attempted—and it surely will be by somebody sometime—it would be cruel not to wish good luck to everyone involved. But the prospect of human cloning causes us grave misgivings. It is physically too risky, it could have untoward effects on the psychology of the cloned child, and in the end we see no medical justification for it. For us, the technology that produced Dolly has far wider significance. As the decades and centuries pass, the science of cloning and the technologies that may flow from it will affect all aspects of human life—the things that people can do, the way we live, even, if we choose, the kinds of people we are. Those future technologies will offer our successors a degree of control over life's processes that will come effectively to seem absolute. Until the birth of Dolly, scientists were apt to declare that this or that procedure would be "biologically impossible"—but now that expression, biologically impossible, seems to have lost all meaning. In the twenty-first century and beyond, human ambition will be bound only by the laws of physics, the rules of logic, and our descendants' own sense of right and wrong. Truly, Dolly has taken us into the age of biological control.

Dolly is not our only cloned sheep. Megan and Morag were our first outstanding successes—Welsh Mountain ewes cloned from cultured embryo cells. Taffy and Tweed, two Welsh Black rams, were cloned from cultured fetal cells at the same time as Dolly and are at least as important as she is, since fetal cells may well be the best kind to work with. If it hadn't been for Dolly, Taffy and Tweed would now be the most famous sheep in the world. At the same time as Dolly, too, we cloned Cedric, Cecil, Cyril, and Tuppence from cultured embryo cells—four young Dorset rams who are genetically identical to one another and yet are very different in size and temperament, showing emphatically that an animal's genes do *not* "determine" every detail of its physique and personality. This is one of several reasons "resurrection" of lost loved ones, human or otherwise, is not feasible.

Ambitions for Genetic Engineering

But Keith and I did not set out simply to produce genetic replicas of existing animals. Some other biologists who have contributed enormously to the science and technology of cloning have indeed been motivated largely by the desire to replicate outstanding—"elite"—livestock. Our broader and longer-term ambitions at Roslin, together with our collaborating biotech company PPL, lie in genetic engineering: the genetic "transformation" of animals and of isolated animal and human tissues and cells, for a myriad of purposes in medicine, agriculture, conservation, and pure science. Future possibilities will in principle be limited only by human imagination. A hint of what might come is provided not so much by Megan and Morag or by Dolly and her contemporaries, who have all been cloned but have not been genetically altered, but by Polly, born the year after Dolly, in 1997.

Polly is both cloned *and* genetically transformed.

Indeed, we should not see cloning as an isolated technology, single-mindedly directed at replication of livestock or of people. It is the third player in a trio of modern biotechnologies [genetic engineering, genomics, and cloning] that have arisen since the early 1970s. Each of the three, taken alone, is striking; but taken together, they propel humanity into a new age—as significant, as time will tell, as our forebears' transition into the age of steam, or of radio, or of nuclear power.

Doubts About Dolly

By Caroline Daniel

A year after Dolly's birth, two researchers questioned her authenticity, Caroline Daniel reports in the following selection. One microbiologist called Dolly's creation "bad science," claiming that the published research failed to show that Dolly and her mother had an identical genetic makeup. In addition, the scientists pointed out that no one in the following year was able to repeat the success, something that must be done to validate scientific experiments. According to Daniel, another criticism that the doubting scientists made was that although Roslin Institute, where Dolly was cloned, says the sheep was cloned from adult mammary cells, the two researchers say that this is not particularly useful knowledge because the institute failed to document whether the cells were fully differentiated or less differentiated. To further cloning research, Daniel explains, scientists need to know what kind of adult cells to use. Moreover, many scientists say that adult cell cloning would not be feasible in the future because it is too expensive and there are not enough commercial applications for it. Caroline Daniel is a journalist who writes for the *New Statesman*, a weekly British magazine.

One year on from science's cloning breakthrough [announced in February 1997], the wonder sheep is looking a bit tatty. How far have we come, genetically, in twelve months?

It's a year since Dolly the sheep trotted demurely on to the media stage, the first mammal to have been cloned from an adult cell. Her birth was hailed by the press: the *Washington Post* ebulliently called her "the biggest story of the year, maybe of the decade, or even the century", and she topped the list of scientific break-

Caroline Daniel, "Not Many Happy Returns, Dolly," *New Statesman*, vol. 127, February 1998, p. 28.

throughs for 1997 in the prestigious US journal *Science*. It has taken a year for the voices of dissent to creep in, but some scientists are, belatedly, starting to ask if Dolly measures up to the hype.

In a letter published recently in *Science*, two researchers, Norton Zinder and Vittoria Sgaramella, question the credibility of the Dolly experiment carried out by Ian Wilmut at Edinburgh's Roslin Institute, in collaboration with PPL [Pharmaceutical Proteins Ltd.] Therapeutics. "It's bad science. I don't know why *Nature* even published it. It's a bad paper by my scientific standards," growls Zinder, a veteran microbiologist from Rockefeller University and a former head of the advisory committee on the international Human Genome Project.

Why has he taken so long to say so? "I've made over 45 attempts to write opinion pieces in various American journals. So far all those who anointed Dolly in the States have ignored me. That's what I was afraid of. I felt this message would never get out amid the media furore," Zinder explains.

A Cry of "Fraud"

At the heart of this criticism, the cry of "fraud" is detectable: the published research, it is claimed, fails to show that Dolly and her "mother" had an identical genetic make-up. The nucleii that were used to create Dolly were taken from a batch of adult mammary cells (hence her name, after Dolly Parton's famous assets) and placed in unfertilised egg cells in a process known as nuclear transfer. The cells had been frozen for three years; their original provenance is now mutton.

The critics' doubts are underlined by the fact that no one in the past year has been able to repeat the experiment. Dolly, though a clone, remains unique. Thus Zinder and Sgaramella write: "It is a well-known tenet of science that a single observation is not to be codified until confirmed by someone in some way. . . . It is the lack of any confirmation that provokes our scepticism."

It is hardly surprising that they want further proof of Dolly's authenticity. No previous embryo cloned from an adult mammal cell had survived more than a few days into its development. The closest anyone came to creating a clone from an adult cell was in 1975 when scientists, using skin cells taken from an adult frog, created tadpoles with heart-beats. Very brief heart-beats.

Which is not to say that scientists haven't successfully created

clones before. Previously, however, they've used foetal material:
cells taken from very early embryos rather than adult cells. Nuclear transfer was first carried out in 1952 by Robert Briggs and
Thomas King, who succeeded in creating an embryo by removing the nucleus from a body cell, putting it into an unfertilised egg
cell whose nucleus had been removed, and then implanting it inside a surrogate mother to grow as a normal embryo. The organism that results is a clone because it has the same genetic material as the original body cell.

What Cells Can Be Cloned?

Once cloning was proved possible scientific attention turned to
the fascinating, sci-fi question of what sort of cells you can use to
make a clone. Can you create one from a piece of Queen Victo-
ria's old hair in a museum? Or from cells scraped away from the
fingertips? Or could it work only with very young cells?

To answer this, it helps to know how cells develop. Normally after an egg and sperm have fused together, an embryo has two cells.
Each divides, to give four, and then these divide again to make
eight. At this stage, each cell still has its own potential to become
an embryo unto itself; in science jargon, they are totipotent. As the
embryo continues to divide, however, the cells begin to differentiate, literally to become different. Some turn into skin cells, others
into blood cells and so on. As cells specialise they lose their ability to generate other types of cell: liver cells can make only liver
cells. A few cells, however, called stem cells, are more like per-
petual dilettantes, not committing to any specialism, but retaining
the ability to generate a variety of cells throughout their lives.

Until the 1960s scientists assumed that only the early totipotent
cells could produce clones. But then research on frogs suggested
otherwise: a nucleus from a specialised cell could be returned to
an undifferentiated state, reprogrammed to start all over again to
make a new, complete organism. In 1966 researchers created adult
frogs from cells taken from tadpole intestines. In the 1980s the
techniques of nuclear transfer were successfully used on mammals. So scientists have been able to prove that some differentiated cells are capable of creating a whole new organism.

But cells, it seems, age and weaken; when they are grown in
culture they divide for a while, but eventually stop reproducing
and die. The older the cell, the fewer times it divides in culture. In

general this is borne out by cloning research: the older the donor cell from which the nucleus is transplanted, the more likely it is that the resulting embryo develops abnormally. That's why Dolly was such a breakthrough.

She raises for the first time the prospect of cloning animals with a proven superior genetic make-up, such as ones that produce large quantities of milk, or particularly fine racehorses (imagine a race run between ten Red Rums). Some have also proposed that cloning from adult cells might help to save pandas and other endangered species from extinction.

Unhappily for the scientific researchers Dolly, rather surprisingly, sheds no light on the important matter of which kind of adult cells are best suited to cloning. Roslin says she was created from an adult mammary cell. But, as Zinder and others point out, we don't know if the cell was a fully differentiated mammary cell or a less differentiated stem cell.

Without this kind of refining knowledge, it's easy to dismiss Dolly as a fluke. Zinder and Sgaramella go so far as to call her "an anecdote, not a result". They point out that the Roslin team started with 434 sheep eggs: 277 fused successfully with transplanted donor nucleii; and, of these, 29 embryos lived long enough to be transferred to a flock of surrogate mothers. Dolly was the sole survivor.

Failure to Clone Adult Cells

James Robl, a university professor and co-founder of Advanced Cell Technology, explains the problem. "It's much harder to make adult cells develop in vitro to a size where they are big enough to implant. They're much less successful than foetal cells. Then there are losses when embryos are transferred. Very few last beyond 40 days."

His firm has attempted to clone from an adult cell. "At this point we have only got one cow, which has got to the second trimester, left." He won't say when it's due.[1]

Other firms' efforts in the past year [1997], have had no better results. Michael Bishop, vice-president of product development for American Breeder Services, says his firm tried. There were some pregnancies, but none survived. A team led by Neal First, professor of animal science at the University of Wisconsin, took

---

1. Advanced Cell Technology produced embryos from adult cells, and twenty-four calves survived to become cows.

adult cell nucleii from rats, monkeys, sheep and pigs and put them into cow eggs to see if they would grow. They did. "We made a few attempts to get pregnancies but there were early losses," he says. "The only way a university would be able to reproduce the research would be if it became more efficient."

The bottom line here is that, despite all the excitement about the prospects for adult cloning, there are simply not enough prospective commercial applications to make it worthwhile for Roslin to repeat its costly research. Robert Foote, a professor of animal physiology at Cornell University, says: "The people in *Science* are being a bit unreal to expect them to repeat it. They can't go on doing research for ever; they need to generate commercial products."

"Pharm" Animals

Roslin's attention now is precisely fixed on how to use some of the techniques that created Dolly to generate cash. Their aim is to "pharm" transgenic animals—those with a foreign gene implanted in them—for use in making pharmaceutical products. This technique brought us the pigs created by Imutran for use in human transplant surgery. They contain an extra human gene that the firm hopes will decrease the chances of pig-organ rejection by humans.

Roslin's most recent modified animal was wheeled out last July [1997]: Polly the sheep. Ian Wilmut described Polly as "the first major step towards the commercial exploitation of nuclear transfer". Polly is different from Dolly in two important ways. She was cloned using foetal, not adult, cells. More importantly she is transgenic. Her extra human gene should mean that when Polly produces milk she will also produce a special protein (in this case one that helps blood clotting), which can be isolated and used for pharmaceutical purposes. The firm hopes, literally, to be able to milk her for cash.

Other companies are busily developing similar products. In January [1998] Advanced Cell Technology showed off its new transgenic cash-calves, George and Charlie, created from foetal cells. They, too, have an additional human gene, intended to make them produce human serum albumin (used in blood transfusions) in their milk.

At American Breeder Services, Bishop says: "Most of our focus is on using foetal cells to try to create transgenic cell-lines for the healthcare industry."

In the area of transgenics, at least, foetal cells are far easier and more commercially viable to manipulate. "Though Dolly helped, what is much more significant is the ability, as Polly confirms, to identify a cell type that can develop well in vitro, and can be genetically manipulated. These techniques deserve much more attention than adult cloning," Robl says.

Dolly's Significance

Even if we have to wait years for a repeat of Dolly, it would be wrong to dismiss her as an irrelevance or even an anecdote. It's true that for researchers she is a bit of a cul-de-sac, but equally she has helped to draw attention to transgenic research, and so boosted the cash being thrown into the field. "It's a big understatement to say that there is much more commercial interest now in this area of research," Robl confirms.

She also provides the most persuasive evidence to date that adult cells can be rejuvenated. This has helped attract more money into research into how cells age and differentiate, which may, in turn, eventually offer insights into the origins of cancer or lead to improvements in grafts, such as growing bone-marrow cells for transplant surgery.

But for the foreseeable future you can forget about human cloning from adult cells. Without more animal research it is unlikely that research using human eggs would be permitted.

Besides which, unless and until the techniques become more efficient, there is no way researchers involved in human embryo research would want to use scarce eggs for experiments. Wilmut used about 1,000 eggs to create Dolly. Compare this with the 8,000– 10,000 eggs available each year for research in Britain, or the waiting lists for eggs for IVF [in vitro fertilization] (between six months and three years). Dr Richard Seed, the American scientist who claimed in January [1998] that he intended to clone infertile people . . . is guilty of hubris, if not pure fantasy, for brushing aside these daunting technical and practical obstacles.

Dolly Is Euthanized

By Justin Gillis

The famous sheep, Dolly, being seriously afflicted with a virus causing incurable lung cancer in sheep, was put to sleep in February 2003. Scientists made the decision to end Dolly's life to save her from a slow, uncomfortable death. In addition to having the virus, Dolly was overweight due to eating from the hands of her many zealous visitors and had contracted arthritis the year before. Although initial studies indicated that her viral disease had nothing to do with being a clone, scientists are still uncertain as to the origins of the other maladies Dolly suffered. After being sent to a taxidermist, Dolly will be put on display at an Edinburgh museum by proud Scottish fans. Justin Gillis is a staff writer for the *Washington Post*, a daily newspaper based in Washington, D.C.

Dolly, the friendly but spoiled sheep whose birth by cloning revolutionized the world's understanding of molecular biology, was put to death [on February 14, 2003] after developing a severe lung problem. She was 6, only middle-aged for a sheep, and her death raised fresh questions about the safety of cloning techniques.

In her brief life, Dolly awed veteran scientists, inspired schoolchildren to study biology, touched off worldwide debate about the morality of tinkering with life, provoked legislation in dozens of countries, and set off a political argument in Congress that will now outlive her.

Her understanding of all this was limited, naturally, but she was not oblivious to her special role as the world's first clone. She would run to meet photographers toting cameras, put her feet up on a fence, pose for pictures and demand food for her trouble.

Like the human for whom she was named, Dolly Parton, she was a diva and a star.

And she suffered for her fame. Dolly ate so much from the hands of visitors that she got fat, even by sheep standards. She contracted arthritis [in 2002], at a young age for a sheep, though it was never clear whether this was from too much picture-posing or from the circumstances of her birth. Dolly suffered other maladies that may, or may not, have been a consequence of her origin.

The Roslin Institute in Edinburgh, Scotland, where Dolly was created and lived her life, announced she had been put to death . . . to spare her further suffering. A virus that causes incurable lung cancer in sheep has been spreading at the institute, killing both cloned and normal animals. A lung scan . . . revealed she was seriously afflicted and, without intervention, would die a slow, uncomfortable death.

"Frankly, the vets were surprised how much the disease had progressed, because she had been living a normal life," said Ian Wilmut, the scientist primarily responsible for creating Dolly, in an interview from Scotland. "It was quite clear things were going to get worse, and this was the kindest thing to do."

Dolly's Death Was Evidently Not Related to Her Cloning

Preliminary findings from a necropsy, the animal version of an autopsy, showed no abnormalities in tissues other than Dolly's lung, Wilmut said. More detailed molecular studies are scheduled to try to learn more about her biology and whether being a clone hastened her death. Wilmut, while acknowledging that many cloned animals have suffered health problems, said Dolly's clone status apparently had no bearing on her demise.

But because of that status, Dolly will go down in history as a scientific and cultural milestone in humanity's efforts to master—and manipulate—the molecular details of life.

She was born in 1996 and unveiled to the world in 1997. For decades before, scientists had speculated that it might be possible to copy an adult animal by using a nucleus from one of its cells, and allow the genetic material inside to direct the development of a near-perfect copy of the adult. The idea had been a staple of science fiction for generations.

Wilmut worked on the problem for years, not to create clones for their own sake but to achieve a more efficient process for making many copies of valuable animals, such as creatures engineered to grow and donate organs to humans. The technique that finally worked was to suck the nucleus out of an adult cell, use it to replace the nucleus of an unfertilized egg, and allow compounds present in the egg to reset the adult nucleus to an embryonic state. That "embryo" was then transplanted to the uterus of a surrogate mother sheep. It took 247 attempts by this method to produce Dolly.

When they finally got cloning to work, Wilmut and his collaborators scheduled a modest announcement, not quite anticipating that the world would immediately imagine human clones. The Roslin Institute was quickly overwhelmed by a media frenzy.

"No one thought this was possible. People had been trying for a long time," said Simon Best, a scientist in Scotland. "The fact that it actually happened was absolutely astonishing."

Under close questioning in 1997, Wilmut revealed that Dolly, whose genes came from a mammary cell of a Finn Dorset ewe, had been named for Parton. The entertainer said she was honored by the tribute. "There's no such thing as baa-aa-aad publicity," she declared.

Human Cloning Is Too Risky

After Dolly was born, scientists quickly cloned mice and then various other mammals. It became clear that human cloning was at least theoretically possible, but also that it would be dangerous. Many cloned animals died in the womb, and others suffered gross abnormalities. A worldwide consensus developed among scientists that human cloning, whatever the ethics of it in principle, was far too risky to attempt in practice.

Many countries have since banned human reproductive cloning, while generally permitting cloning research on microscopic human cells. Cloning a sick person's cells might be one way to create tissues that could replace ailing kidneys, livers or other organs. The United States has passed no law, as Congress deadlocked between a camp that wants to ban only reproductive cloning and another that wants to ban research cloning, as well. Fringe groups have claimed success at cloning people but have produced no evidence.

Meantime, the animal research has proceeded apace, with several American companies creating clones for commercial purposes. The Food and Drug Administration is trying to decide whether milk from clones or meat from their offspring should be allowed into the food supply.

Coupled with new findings involving immature cells called stem cells, the research that started with Dolly has shown that animal cells are far easier to manipulate than scientists once thought. They have had to throw out their textbooks on biological development and start over.

Dolly gave birth, by normal sexual reproduction, to six lambs, all of which appeared normal. Even in death, the world's most exalted sheep is destined for continuing stardom. A focus of intense pride in Scotland, Dolly will be sent to a taxidermist, stuffed and put on display at a museum in the center of Edinburgh.

CHAPTER 3

Current Cloning Controversies

Should Humans Be Cloned? A Symposium

Part I: Gregory E. Pence; Part II: Michael A. Goldman

In the following symposium two professors debate the question of whether human cloning should be legal when and if the process becomes a reality. On the one hand, Gregory E. Pence says that cloning is a legitimate way for people to have children. Pence believes that prejudice and ignorance, including the idea that a cloned child would lack human dignity because he or she would not be unique, have created unnecessary barriers to human cloning. Pence counters such beliefs by explaining that the cloned child would have his or her own memory, character, and family, which would assure the child's individuality. Pence believes the strongest reason to support human cloning is to prevent terrible genetic diseases passed on from parents to children. In addition, he points out that the federal government would have no constitutional right to prohibit parents from having a cloned child.

On the other hand, Michael A. Goldman argues against human cloning because of the medical and psychological risks involved. For one thing, Goldman says, cloning circumvents the normal process of cell division that occurs during sexual reproduction, which could result in genetic defects, some of which would not appear until years later. Another problem is that the emotional well-being of the cloned child could be shattered if the child doubted his or her individuality and autonomy. Goldman points out that infertile couples have other options, including adoption, available to them in lieu of cloning. For these and other reasons, Goldman concludes that scientists and society should proceed with caution in the area of human cloning.

Gregory E. Pence is a professor in the school of medicine and the philosophy department at the University of Alabama, Birmingham, and author of *Who's Afraid of Human Cloning?* Michael A. Goldman is a biology professor at San Francisco State University and has written for the *New York Times* and the *Wall Street Journal*.

Part I

Q : If human cloning becomes a reality, should it be a legal option? Gregory E. Pence: *Yes: Cloning could become a legitimate way for some people to have children.*

When we watch the film *Snow Falling on Cedars*, we like to think that we would have defended persecuted Japanese-Americans. Had we lived in Alabama in Martin Luther King Jr.'s time, we like to think we would have marched with him to end racial segregation. Fighting prejudice is easy over a battle long won.

It is less easy to fight prejudice in the present or even to see it in one's world. Yet prejudice surrounds discussions of originating a person by cloning, and few people have risen to fight it. Consider the many commentators who talk about "whether the clone" would be mistreated. "Clone" is so pejorative that its very use indicates question-begging attitudes, like the introduction of a report about equal rights for women that begins, "The chicks say. . . ."

We are in a similar situation now about cloning humans. Much prejudice and ignorance abound, abetted by movies such as *Multiplicity, Blade Runner* and, in an even worse case, the movie starring Arnold Schwarzenegger, *The 6th Day*. (Schwarzenegger goes away from home to discover he has been replaced by an identical copy himself complete with memories and habits. Even Schwarzenegger's wife can't tell them apart. But who is the original and who is the clone? People who don't know much about cloning want to know.)

Most objections to human cloning do not focus on possible physical harm to the new child but instead raise psychological, religious, social and ethical objections, where prejudice and ignorance are seen.

The Report on Cloning by the National Bioethics Advisory Commission, or NBAC, mentioned that a "massive majority" of Americans fear human cloning. Other critics say that "almost everybody" thinks it's a bad idea and that opinion polls are "nine to one" against cloning.

This is just the ad populum fallacy. Most people wanted to inter Japanese-Americans during World War II, most Americans accepted segregation before the civil-rights movement, and too many people today still think that it's bad to be gay or lesbian. Most people 20 years ago feared "test-tube babies" and "genetic-engineering therapy" before these were understood.

Cloning Does Not Undermine Human Dignity

A common objection is that originating a person by cloning will undermine human dignity because the new child will not be unique but a copy of an ancestor. This objection is wrong on three counts. First, the mitochondria of the egg in which the ancestor's nucleus is housed will contribute 1 percent of the genes, so the new child will be only 99 percent genetically the same. Second, only the ancestor's genes get copied, not the person who was formed with memories, character, free will and a family. Finally, the objection assumes that the only way a person can be morally valuable is if her genome is unique. So identical twins and triplets lack moral value?

Moral value comes not from a random spin of the genetic roulette wheel but from how society creates moral rules about how to treat humans. Such moral rules must be grounded on morally relevant characteristics of beings, such as feeling pain and having consciousness, and not on skin color, sexual orientation or method of origination.

"But the child will be harmed because it will know that it was not wanted for itself but for the characteristics of the ancestor." So goes another common objection. According to the Alan Guttmacher Institute, two-thirds of American women will be unintentionally pregnant at some time during their lives and many of them will not get an abortion. So many American babies are still not planned, not "wanted for themselves."

A child created by cloning would not need to know how or why it was created. Critics assume it would know, but critics in the past assumed similar things about "test-tube babies"—assuming, of course, that it was a terrible thing to reveal to an adult the ordeals and sacrifices of money that the infertile couple went through to create the child.

But let's bite the bullet and suppose the adolescent is told the truth—that is, he's told that a certain genotype seemed desirable to the parents. "Your Uncle Frank was a great guy and never had a sick day in his life. He loved skiing, played the saxophone and graduated at the top of his class at West Point." Would knowing that kind of information harm a teen-ager?

Before you answer, let me tell you how millions of children were created in most of history. My own grandmother, Vernie Burner, was one of 10 children born to a Mennonite Shenandoah Valley farmer who needed a large family to help him cut hay, pick apples and feed the chickens. Back then, children were created to be farmhands. Parents didn't save for their college education, take them to gymnastic classes or spend every night at soccer games.

Ironically, when social critics of today look back to our American past, it is precisely this era they see through rose-colored glasses as a time of great American character, true moral values and family strength. Regardless of whether they are right, most children back then were not "created because they were wanted for themselves."

Another example may put cloning in perspective. Modern societies could do something right now that would prevent great harm to millions of children. Evidence for this harm is incontestable and admitted by virtually everyone. This is the harm of allowing women to smoke and drink while pregnant. Incidence of cleft palate recently was shown in the *New England Journal of Medicine* to vary directly with the number of packs smoked per day by the mother. If we chose, we could make it illegal for a woman to do so as soon as she learned she was pregnant. But we won't so choose, and not just because of the American Civil Liberties Union. We wrongly think such harms to children are normal and that nothing really can be done to stop them. Instead we focus our fears, novels and science-fiction movies on the sensationalized harm of cloning.

The geneticist J.S. Haldane once quipped that in such debates, "An ounce of arithmetic is worth a ton of verbal argument." He was right. In fact, originating humans by cloning requires in vitro fertilization, which is expensive ($8,000 per attempt) and inefficient (only 20 in 100 couples ever take home a baby created this way). In the United States during the last 20 years, 50,000 babies have been created through assisted reproduction. If the same figure holds for the future, and when cloning becomes safe, maybe 500 babies will be originated by asexual reproduction of the genome of some ancestor. And all embryos created this way need to be gestated for nine months by a real woman, who might miscarry or change her mind and abort.

So a small number of people will be created by cloning and, when they are, we will have evidence to think we can do so without physically harming such children. Rather than convene national commissions to inveigh against such imaginary evils, we might do better to form them to combat the real ways children are born harmed every day.

Constitutionally, there is no basis for the NBAC or the federal government to tell married couples when and how to have children. Nothing gives the U.S. government this power in the Constitution or Bill of Rights. The only possible basis is the Interstate Commerce Act. Even if sperm and eggs occasionally cross state lines in being shipped to infertility clinics, it is preposterous to think that such conditions were what Congress intended in creating the act, much less that such an interpretation would pass review by the Supreme Court.

But is there any good reason why a couple would want to originate a child by cloning? First, notice how hypocritical we are in putting cloning to this test and not the teen-age couple next door. If each of our parents had to pass a "good-reasons" test before some governmental committee, many of us would not have been born. Second, what counts as a "good reason" will vary with all the different conceptions of the good life that are found in philosophy. Many people now have children so that "something of me lives after my death" or so "someone will take care of me when I'm old." These reasons are self-interested.

There are two great, positive arguments why human cloning should not be legally banned as a medical option. One day soon, after studies on chimpanzees and monkeys prove that cloning will be safe for resulting children, these arguments will become important.

Cloning Could Prevent Terrible Genetic Disease

The best argument is to prevent terrible genetic disease. Consider an Orthodox Jewish couple who had three children die of Tay-Sachs disease. They are reluctant to conceive again and to risk another child dying of this disease. Having children is very important to them, but abortion of embryos or using a surrogate mother are not options. Using cloning, the father's genotype could be inserted into the egg from another woman and the wife could gestate that embryo. In this way, a lineage going back a thousand years could be preserved and continued.

Consider also a man who is azoospermatic and his wife, who has no viable eggs. They could create two children, each a copy of one parent's genotype, and have the wife gestate each. In this way, they could have a genetic connection to two children.

In public policy, we think it good that parents are biologically bonded to their children. If so, why isn't this permissible?

We should be very careful before we ban a possible medical option. Twenty years ago, Congress put a ban on federal funds used in any research involving human embryos and we now struggle to lift it (and perhaps we will not be successful). One hundred years ago, a brave Alabama physician named J. Marion Sims first inseminated a woman with her husband's sperm to overcome infertility. Sims produced a pregnancy but universally was condemned and had to stop. In the 1930s, another physician tried this procedure again but also was forced to stop. Not until the 1960s did society accept this medical option. As a result, thousands of couples over a century were denied children they might have had. In a situation of rapidly changing technology, like the early days of AIDS, it is premature to criminalize a possible, future medical option.

Part II

Michael A. Goldman: *No: We do not know enough about the medical and psychological risks involved.*

The advent of cloning animals from adult cells, and the possible application of this technique to humans,have engendered much debate. In combination with recombinant DNA technology [genetic engineering] and embryonic stem-cell methodology, cloning is ushering in a staggering new millennium in medicine. I strongly advocate the experimental use of very early embryos (less than two weeks, long before the nervous system is formed) for production of embryonic stem cells and for conducting basic research on the early development of humans. However, there are no scientific grounds for pursuing the use of cloning to produce a human child.

Cloning by somatic nuclear transfer is the introduction of the nucleus of a cell obtained from an existing individual into an egg cell (oocyte) from which the original nucleus has been removed. The egg subsequently begins cell division, under the direction of the introduced nucleus, producing a multicellular embryo that is the genetic twin, or clone, of the nuclear donor. From this very early embryo, scientists can produce embryonic stem cells, isolate cells to learn about genetics and development or attempt to rear an individual to adulthood. Carried to its end point, this is the technology that produced Dolly the sheep, an achievement that astounded scientists and the public alike. But the use of this technology to produce a human child is both undesirable and unreasonable at this time.

First, consider the reasons for producing a cloned human. We may think that a particular individual is exceptionally talented, or exceptionally pretty, and should be perpetuated. We may have lost a loved one and strongly desire to replace this person. Such thinking is fueled by a serious misunderstanding of what it means to be a genetic clone. It is very clear that an individual is a product of its environment as well as its genetic makeup and that a cloned individual may differ quite significantly from the "prototype." In addition, a clone would not share the same cytoplasmic factors, such as mitochondrial DNA, unless the egg donor and the prototype were the same. Thus, the idea that we can duplicate a person by producing a genetic clone is biologically unsound.

Alternatives to Cloning

Treatment of infertility is one application that deserves serious consideration as a legitimate reason for wanting to produce a clone. A couple unable to conceive a child may want a child genetically related to, or identical to, at least one of them. The psychological imperative to pass on one's genes is a sensitive issue. But that imperative is to produce children who have half of our own genetic material, not children who are genetically identical to us. If infertility is in part genetically based, we may be passing on the infertility problem to a new generation. Fertility treatments abound which also are alternatives to cloning. There are strong arguments for adopting the many children who do not have loving families, rather than using extreme measures to produce genetically identical children.

Secondly, we must consider the practicality of producing a cloned human. While technological advances are likely, our present capabilities make the process remarkably inefficient. In cloning Dolly, Ian Wilmut and colleagues, after many years of related experiments, used about 400 eggs. Of these, about 277 actually fused with the donor nucleus. From these couplets, 29 reached a stage appropriate for implantation into 13 foster mothers, and only one—Dolly—survived long enough for media coverage. Success rates with some mammals, such as mice, have been considerably higher, but none have reached a level we would consider impressive for an optional medical procedure. Different animals, including humans, are likely to respond differently to the procedures and, paraphrasing what they say about mileage in the auto industry, "your success rate may even differ."

The Dangers the Cloned Child Might Face

Thirdly, we should take account of the potential dangers a cloned human might face. As with any medical procedure, there is risk. While a certain frequency of miscarriage or birth defects occurs normally, a slightly elevated rate because of the manipulations involved in cloning would not be surprising. And we cannot rule out that there might be a drastic increase in these complications as a result of cloning. Cloning subverts the normal process of meiosis, a type of cell division used in the production of egg and sperm. During this process, chromosomes and genes are shuffled so that new, unique combinations of genetic characteristics result. More importantly, major errors in the DNA are filtered out as damaged cells fail to complete the process. Thus, the eggs and the sperm that fertilize them have passed a strict quality-control test. Somatic cell nuclei, used in cloning, have not only skipped this step, but have been replicating in the body or in culture, accumulating errors with time. We also know that the genetic material is differentially "marked" by the male and female germlines. Some genes only are expressed if inherited from the father, while other genes only are expressed when inherited from the mother, a phenomenon known as genomic imprinting. It is possible that abnormal gene expression might be seen in cloned embryos. The problems outlined here are possible reasons that the efficiency of cloning is relatively low, and this might be a biological barrier that we will never be able to cross.

The most exciting thing that scientists learned about developmental biology from cloning Dolly might also be one of the most ominous. Somatic cells—those we see in adults—are said to be "differentiated." Somatic cells have specialized to express only a small subset of their 50,000–100,000 genes. Before Dolly was cloned, we thought that this differentiation process was irreversible—once differentiated, a nucleus could not go back and produce a wide variety of other types of cells. Further, adult somatic cells do not divide as efficiently as the cells of early embryos. The cultured breast epithelial cells that were the source of the nucleus that produced Dolly showed us that these two problems do not present an absolute barrier to cloning. But we cannot rule out the possibility that the one in 277 or so fused cells that were successful just happened to be a little atypical. These cells might have been less specialized or more capable of cell division than the rest. Cancerous or precancerous cells may differ from normal cells in their production of certain specialized gene products and in their capacity to undergo continued cell division. Dolly seems quite healthy, but she could have an elevated risk of cancer. The first cloned human might come from an exceptional cell, leaving the newborn on the brink of this grave affliction.

Complications Might Appear Later in Life

In addition, there could be complications that would appear only later in life. Human cells are limited to about 50 cell divisions (the "Hayflick number"), stalked by a relentless internal clock that operates in part as the chipping away of the ends of chromosomes called telomeres. The clock is reset in the process of sexual reproduction. But Dolly skipped this process—she was cloned from a somatic cell rather than from the union of egg and sperm. Scientists have shown that her telomeres are shorter than those of similar-aged ewes, suggesting that she might age prematurely. In contrast, a group of cloned calves had telomeres that were longer than usual, suggesting the possibility of an extended life span. But longer telomeres are also sometimes seen in cancer cells. Do parents want to risk, knowingly, having a child whose life span will be limited because of the fertility treatment they undertook?

Finally, we should consider the emotional well-being of the cloned child. If a child were to learn that he had been cloned for specific purposes, such as the ability to play the piano, he might face unreasonable expectations. He might question his existence as an individual, his autonomy. Was he a product manufactured to a specific end, with no rights to make his own decisions? If the prototype had died young of a disease, or even an accident, would the clone think his own fate was sealed? Might the prototype's spouse fall in love with a younger clone? While some of these problems also are faced by identical twins, the latter are not the products of medical intervention.

I want to emphasize that applications of cloning and related technologies can and should be used to understand and alleviate illness, genetic disease, cancer and the ravages of aging.

Stem cells are cells that are capable of perpetuating by cell division, or of giving rise to a variety of specialized cell types. Our ability to work with embryonic stem cells is one of the most dramatic advances of the last decade. Until very recently, use of federal funds in this essential research has been prohibited because it involves the manipulation of cells from human embryos.

In August 2000 the National Institutes of Health issued guidelines that will permit federally funded scientists to work with embryonic stem cells produced in the laboratories of privately funded scientists. (The stem cells are taken from embryos that were being discarded as a by-product of in vitro fertilization.) While these new guidelines dramatically reverse earlier tight restrictions, the scope of federally funded research will have to be broadened before we can realize the full potential of this research in the public sector. The ability to provide matched tissues for transplantation will require the production of a cloned early embryo for each intended recipient. In other words, before a given individual can receive a tissue transplant there must be a cloned early embryo from one of their cells. However, progress in basic research on gene expression may one day allow us to accomplish similar feats with adult stem cells.

Our analysis must remain theoretical, as we do not yet have data on cloned humans, and the information on other cloned organisms still is very limited. It is just this lack of information, this uncertainty, that should make us proceed with caution. Rightly or wrongly, we do not use high-risk, unproven medical treatments, even in terminal cases. Our first concern in medical intervention is to "do no harm." Why should we be so quick to use a medical procedure that might threaten the very offspring we seek to produce? The reasons for undertaking human cloning are tenuous at best, and alternative treatments for infertility, including adoption, are available. Cloning to produce a child is without scientific merit and has raised serious ethical concerns. Unless and until circumstances change dramatically, it should not be a legal medical option, and it is not a rational medical option.

The Pros and Cons of Regulating Cloning Technologies

By Gregory Stock and Francis Fukuyama

In March 2002 Gregory Stock and Francis Fukuyama, two important figures in the cloning controversy, participated in an online debate from which the following selection was taken. Stock, director
of the Program of Medicine, Technology, and Society at the University of California at Los Angeles, believes that laws regulating
cloning will not stop humans from undertaking the practice. Moreover, he contends that a ban on cloning would unfairly prevent
people from accessing technologies that could improve their lives.
In addition, because the policies needed to regulate cloning technology would be harsh and intrusive, they would create more harm than
the technology itself would, he claims. Stock argues that unwarranted
laws regulating cloning would delay medical advances and cause unnecessary suffering for those with serious diseases who could have
benefited from the technology.

Fukuyama, professor of international political economy at the Paul
H. Nitze School of Advanced International Studies of Johns Hopkins University, disagrees with Stock on the safety of cloning, pointing out that genetic modification is highly complex. He argues that
there could be long-term, irreversible genetic effects that might show
up in cloned children decades later. Fukuyama also worries that if
cloning and other reproductive technologies are not regulated, people
will start "designing" children to have positive traits such as as-
sertiveness or to lack so-called negative traits such as dark skin pig-

Gregory Stock and Francis Fukuyama, "The Clone Wars," *Reason*, vol. 34, June 2002, p. 34.
Copyright © 2002 by Reason Foundation, 3415 S. Sepulveda Blvd., Suite 400, Los Angeles, CA
90034, www.reason.com. Reproduced with permission.

mentation. Unlike Stock, Fukuyama believes that regulations can work because similar laws have worked in the past.

What is the case for allowing genetic and other biological manipulations with the potential to change human beings? As stem cell research, cloning, and other technologies develop, perhaps no other question is more central to our future as a species—and perhaps no other question is as hotly contested.

In the wake of the first meeting of the President's Council on Bioethics and as Congress considered new legislation on the matter, we invited two of the major players in the field to debate the issue on *Reason* [magazine's] Web site. Gregory Stock, who makes the affirmative case, is director of the Program of Medicine, Technology, and Society at the University of California at Los Angeles School of Medicine. He is also the author of the new book *Redesigning Humans: Our Inevitable Genetic Future.* Arguing against genetic and biological manipulations is Francis Fukuyama, professor of international political economy at the Paul H. Nitze School of Advanced International Studies of Johns Hopkins University. He is the author of *Our Posthuman Future: Consequences of the Biotechnology Revolution.* . . .

Gregory Stock: There has been a lot of hand wringing recently about cloning. Considering that not a single viable cloned human embryo has yet been created, that the arrival of a clinical procedure to do so seems quite distant, and that having a delayed identical twin (which is, after all, what a clone is) has limited appeal, why all the fuss?

The fuss arises because cloning has become a proxy for broader fears about the new technologies emerging from our unraveling of human biology. Critics like Francis Fukuyama imagine that if we can stop cloning we can head off possibilities like human enhancement, but they're dreaming. As we decipher our biology and learn to modify it, we are learning to modify ourselves—and we will do so. No laws will stop this.

Embryo selection, for example, is a mere spin-off from widely supported medical research of a sort that leaves no trail and is feasible in thousands of labs throughout the world. Any serious attempt to block such research will simply increase the potential dangers of upcoming technologies by driving the work out of sight, blinding us to early indications of any medical or social problems.

A Ban Would Prevent People from Making Choices

The best reason not to curb interventions that many people see as safe and beneficial, however, is not that such a ban would be dangerous but that it would be wrong. A ban would prevent people from making choices aimed at improving their lives that would hurt no one. Such choices should be allowed. It is hard for me to see how a society that pushes us to stay healthy and vital could justify, for instance, trying to stop people from undergoing a genetic therapy or consuming a drug cocktail aimed at retarding aging. Imposing such a ban requires far more compelling logic than the assertion that we should not play God or that, as Fukuyama has suggested, it is wrong to try to transcend a "natural" human life span.

What's more, a serious effort to block beneficial technologies that might change our natures would require policies so harsh and intrusive that they would cause far greater harm than is feared from the technologies themselves. If the War on Drugs, with its vast resources and sad results, has been unable to block people's access to deleterious substances, the government has no hope of withholding access to technologies that many regard as beneficial. It would be a huge mistake to start down this path, because even without aggressive enforcement, such bans would effectively reserve the technologies for the affluent and privileged. When abortion was illegal in various states, the rich did not suffer; they just traveled to more-permissive locales.

Restricting emerging technologies for screening embryos would feed deep class divisions. Laboratories can now screen a six-cell human embryo by teasing out a single cell, reading its genes, and letting parents use the results to decide whether to implant or discard the embryo. In Germany such screening is criminal. But this doesn't deny the technology to affluent Germans who want it: They take a trip to Brussels or London, where it is legal. As such screenings become easier and more informative, genetic disease could be gradually relegated to society's disadvantaged. We need to start thinking about how to make the tests more, not less, accessible.

But let's cut to the chase. If parents can easily and safely choose embryos, won't they pick ones with predispositions toward various talents and temperaments, or even enhanced performance? Of course. It is too intrusive to have the government second-guessing such decisions. British prohibitions of innocuous choices like the sex of a child are a good example of undesirable government intrusion. Letting parents who strongly desire a girl (or boy) be sure to have one neither injures the resulting child nor causes gender imbalances in Western countries.

Sure, a few interventions will arise that virtually everyone would find troubling, but we can wait until actual problems appear before moving to control them. These coming reproductive technologies are not like nuclear weapons, which can suddenly vaporize large numbers of innocent bystanders. We have the luxury of feeling our way forward, seeing what problems develop, and carefully responding to them.

The Real Danger Is Delaying Medical Advances

The real danger we face today is not that new biological technologies will occasionally cause injury but that opponents will use vague, abstract threats to our values to justify unwarranted political incursions that delay the medical advances growing out of today's basic research. If, out of concern over cloning, the U.S. Congress succeeds in criminalizing embryonic stem cell research that might bring treatments for Alzheimer's disease or diabetes—and Fukuyama lent his name to a petition supporting such laws—there would be real victims: present and future sufferers from those diseases.

We should hasten medical research, not stop it. We are devoting massive resources to the life sciences not out of idle curiosity but in an effort to penetrate our biology and learn to use this knowledge to better our lives. We should press ahead. Of course, the resultant technologies will pose challenges: They stand to revolutionize health care and medicine, transform great swaths of our economy, alter the way we conceive our children, change the way we manage our moods, and even extend our life spans.

The possibilities now emerging will force us to confront the question of what it means to be a human being. But however uneasy these new technologies make us, if we wish to continue to lead the way in shaping the human future we must actively explore them. The challenging question facing us is: Do we have the courage to continue to embrace the possibilities ahead, or will we succumb to our fears and draw back, leaving this exploration to braver souls in other regions of the world?

There Are Good Reasons to Regulate Future Biotechnologies

Francis Fukuyama: Gregory Stock offers two sets of arguments against restricting future biotechnologies: first, that such rules are unnecessary as long as reproductive choices are being made by individual parents rather than states, and second, that they cannot be enforced and will be ineffective even if they were to be enacted. Let me respond to each in turn.

While genetic choices made by parents (either in the short run, via pre-implantation genetic diagnosis, or in the more distant future, through germline engineering) are on the whole likely to be better than those made by coercive states, there are several grounds for not letting individuals have complete freedom of choice in this regard.

The first two are utilitarian. When we get into human germline engineering, in which modifications will be passed on to successive generations, safety problems will multiply exponentially over what we today experience with drug approval. Genetic causation is highly complex, with multiple genes interacting to create one outcome or behavior and single genes having multiple effects. When a long-term genetic effect may not show up for decades after the procedure is administered, parents will risk a multitude of unintended and largely irreversible consequences for their children. This would seem to be a situation calling for strict regulation.

A second utilitarian concern has to do with possible negative externalities, which is the classic ground for state regulation, accepted by even the most orthodox free market economists. An example is sex selection. Today in Asia, as a result of cheap sonograms and abortion, cohorts are being born with extremely lopsided sex ratios—117 boys for every 100 girls in China and at one point 122 boys for every 100 girls in Korea. Sex selection is rational from the standpoint of individual parents, but it imposes costs on society as a whole in terms of the social disruption that a large number of unattached and unmarriageable young males can produce. Similar negative externalities can arise from individual choices to, for example, prolong life at the cost of a lower level of cognitive and physical functioning.

A further set of concerns about the ability to "design" our children has to do with the ambiguity of what constitutes improvement of a human being, particularly when we get into personality traits and emotional makeup. We are the product of a highly complex evolutionary adaptation to our physical and social environment, which has created an equally complex whole human being. Genetic interventions made out of faddishness, political correctness, or simple whim might upset that balance in ways that we scarcely understand—in the interest, for example, of making boys less violent and aggressive, girls more assertive, people more or less competitive, etc. Would an African American's child be "improved" if we could genetically eliminate his or her skin pigmentation?

The final issue concerns human nature itself. Human rights are ultimately derived from human nature. That is, we assign political rights to ourselves based on our understanding of the ways members of our species are similar to one another and different from other species. We are fortunate to be a relatively homogenous species. Earlier views that blacks were not intelligent enough to vote, or that women were too emotional to be granted equal political rights, proved to be empirically false. The final chapter of Greg Stock's book opens up the prospect of a future world in which this human homogeneity splinters, under the impact of genetic engineering, into competing human biological kinds. What kind of politics do we imagine such a splintering will produce? The idea that our present-day tolerant, liberal, democratic order will survive such changes is far fetched: [Philosopher Frederick] Nietzsche, not [philosophers] John Stuart Mill or John Rawls, should be your guide to the politics of such a future.

Cloning Can Be Regulated

Stock's second set of arguments is based on his belief that no one can stop this technology. He is certainly right that if some future biotechnology proves safe, cheap, effective, and highly desirable, government would not be able to stop it and probably should not try. What I am calling for, however, is not a ban on wide swaths of future technology but rather their strict regulation in light of the dangers outlined above.

Today we regulate biomedical technology all the time. People can argue whether that technology is properly regulated and where exactly to draw various regulatory lines.

But the argument that procedures that will be as potentially unsafe and ethically questionable as, say, germline engineering for enhancement purposes cannot in principle be regulated has no basis in past experience.

We slow the progress of science today for all sorts of ethical reasons. Biomedicine could advance much faster if we abolished our rules on human experimentation in clinical trials, as Nazi researchers did, and allowed doctors to deliberately inject infectious substances into their subjects.

Today we enforce rules permitting the therapeutic use of drugs like Ritalin, while prohibiting their use for enhancement or entertainment.

The argument that these technologies will simply move to more favorable jurisdictions if they are banned in any one country may or may not carry weight; it all depends on what they are and what the purpose of the regulation is. I regard a ban on reproductive cloning to be analogous to current legislation banning incest, which is based on a similar mix of safety and ethical considerations. The purpose of such a ban would not be undermined if a few rich people could get themselves cloned outside the country. In any event, the world seems to be moving rather rapidly toward a global ban on reproductive cloning. The fact that the Chinese may not be on board shouldn't carry much weight; the Chinese also involuntarily harvest organs from executed prisoners and are hardly an example we would want to emulate.

I don't think that a set of regulations designed to focus future biomedicine on therapeutic rather than enhancement purposes constitutes oppressive state intervention or goes so far beyond the realm of what is done today that we can declare its final failure in advance. By Greg Stock's reasoning, since rules against doping in athletic competitions don't work 100 percent of the time, we should throw them out altogether and have our athletes compete not on the basis of their natural abilities but on the basis of who has the best pharmacologist. I'd rather watch and participate in competitions of the old-fashioned kind.

Therapeutic Cloning Should Be Supported

By Jerome Groopman

Jerome Groopman disagrees with the 2002 decision of the President's Council on Bioethics to enforce a four-year moratorium on cloning for biomedical research (therapeutic cloning). In therapeutic cloning, clonal embryos are created for the production of stem cells used for researching diseases, in contrast to reproductive cloning, which aims to create a child. In the following viewpoint, Groopman calls the council's decision a "painful disappointment," especially for those who could benefit from the potentially lifesaving research. He dismisses arguments against therapeutic cloning, which postulate that the embryo is a child that deserves full protection. Groopman argues that an early-stage zygote (a fertilized egg), which is terminated during the therapeutic cloning process, cannot be hurt or suffer because it possesses no organs, nervous system, or precursors to a nervous system. Moreover, he points out that the question of when life begins will never be solved because it is a matter of personal belief. Because of this fact, he believes the council's moratorium will not bring public consensus about the ethics of therapeutic cloning but only delay important research. Jerome Groopman is a professor of medicine at Harvard Medical School, chief of experimental medicine at Beth Israel Deaconness Medical Center, and director of the AIDS Oncology Program at Harvard's Cancer Center.

T he President's Council on Bioethics, chaired by Dr. Leon R. Kass, presented its long-awaited report on human cloning to the White House on July 10 [2002]. The council unanimously

advised against "cloning to produce children," commonly called "reproductive cloning." But on "cloning for biomedical research"— therapeutic cloning to produce stem cells to try to ameliorate disease—it split. Of the 17 members, ten (including Kass) voted against it. They couched their rejection as a compromise since they called not for a permanent ban but for a four-year moratorium. This moratorium, according to the letter accompanying the report, would allow "a thorough federal review . . . to clarify the issues and foster a public consensus about how to proceed." It would also give researchers time to seek alternative ways to generate stem cells. But for scientists and, more importantly, for the millions of patients with incurable maladies, the compromise is a painful disappointment. It shackles potentially lifesaving research and provides no clear framework to advance the ethical debate. What's more, the arguments deployed on its behalf don't withstand scrutiny.

"Reproductive" and "therapeutic" cloning both begin by transferring the nucleus of an adult cell into an egg. The egg is then given a jolt of electricity and begins to divide under the direction of the inserted adult nucleus. As in normal sexual reproduction, here the cluster of dividing cells is called a zygote. In therapeutic cloning the process is stopped before two weeks, when the zygote is composed of fewer than 150 cells—approximately the size of the dot on the top of the letter "i" on this page. At that time the embryonic cluster contains numerous stem cells. It is these stem cells that scientists are studying as therapies to regenerate diseased tissues in patients with Parkinson's disease, spinal cord paralysis, and juvenile diabetes.

If the process is not stopped at 14 days, however, and the zygote is later placed in a uterus, the potential exists to produce a baby. Reproductive cloning exploded onto front pages five years ago when British scientists successfully cloned a sheep named Dolly [in 1996]. This success with an ovine species has since been extended to other species, like rodents and cats. From the outset, responsible scientists have insisted that reproductive cloning of Homo sapiens not be pursued. But scientific condemnations don't carry the force of law. And once Americans began debating how best to outlaw reproductive cloning, some began arguing that the only way to do so was to ban therapeutic cloning as well—both because the latter could lead to the former and because the latter morally resembled the former. That's where the President's Council on Bioethics comes in.

The council's majority acknowledged that cloning for biomedical research might cure disease; but they advocated a four-year moratorium, nonetheless, for two basic reasons. Their first argument was that therapeutic cloning involved terminating, in the report's words, a "nascent human being." This is because the egg, which receives the adult nucleus, could eventually be moved from the test tube into the uterus and possibly be carried to term to produce a baby. This potentiality, the majority reasoned, means the zygote should be afforded similar respect as a child. Dr. William Hurlbut, a Stanford University bioethicist and a member of the majority, made this explicit in his personal statement appended to the report:

> Anything short of affirming the inviolability of life across all of its stages from zygote to natural death leads to an instrumental view of human life. . . . The inviolability of human life is the essential foundation on which all other principles of justice are built, and any erosion of this foundation destabilizes the social cooperation that makes possible the benefits of organized society.

Other council members likened the zygote to more familiar types of vulnerable human life, like the disabled or the deformed. As the majority statement put it:

> To be sure, the embryo does not yet have, except in potential, the full range of characteristics that distinguish human species from others, but one need not have those characteristics in evidence in order to belong to the species. And of course human beings at some other stages of development—early in life, late in life, at any stage of life if severely disabled—do not forfeit their humanity simply for want of these distinguishing characteristics.

An early-stage zygote is crucially different from the disabled, the deformed, the fragile young, or the fragile old. Before 14 days—the legal cutoff Britain has established for scientific research—the zygote has developed no organs, no nervous system, nor even the precursor to a nervous system. This absence of the most primitive neural anatomy means that biologically the zygote cannot receive any form of stimulation related to the senses, cannot perceive or cogitate, and thus cannot be hurt or suffer.

But, for the council's majority, this biological fact is irrelevant

"because the embryo's human and individual genetic identity is present from the start." But if potentiality alone conferred sanctity, then the single adult nucleus, which holds the genetic program of the later zygote, would also be worthy of protection. No human cell could be discarded—ever. No biopsy. No tube of blood. No vial of frozen sperm or egg. Absurd? Of course. But this is where the majority's logic leads.

Members of the council's seven-member minority—like Harvard political theorist Michael Sandel and Dartmouth neurobiologist Michael Gazzaniga—pointed out other ways in which currently accepted medical practice and federal law clash with the proposed sanctity of the undifferentiated zygote. For example, between 50 percent and 80 percent of all early zygotes, created naturally by a sperm fertilizing an egg during typical coitus, are spontaneously expelled from a woman's body. By the majority's logic, this represents the death of a nascent human being. Yet such regular events in a woman's life do not trigger grief, let alone funerals. Moreover, a single dividing egg—one zygote—can split to make identical twins until at least 14 days after fertilization. This means we cannot designate a single person within a single fertilized egg, because two individuals can emanate from it. Moreover, after the embryo divides into two zygotes, it can recombine and form a single zygote again. If a zygote were a single person, then the individual born from this recombination event would have to be understood as two people. Did two souls merge back into one?

Indeed, President George W. Bush himself hasn't fully respected the undifferentiated zygote's humanity. In his August 2001 stem-cell compromise, the president said that while no *new* stem cells could be derived from in vitro fertilized (IVF) embryos that would be discarded in the future, stem-cell lines derived from a limited number of *existing* embryos discarded during IVF could be employed in research. But if there is a continuous line of life that begins at the first cell division of the fertilized egg, then using stem cells from IVF embryos is, in essence, exploiting body parts from terminated nascent human beings.

The Slippery Slope of Therapeutic Cloning

The majority's second main argument is the slippery slope: Even if therapeutic cloning isn't fundamentally the same as reproductive cloning, it will lead us toward it—and worse. As the majority writes,

To set foot on this slope is to tempt ourselves to become people
for whom the use of nascent human life as research material becomes routinized and everyday. . . . Today, the demand is for stem
cells; tomorrow it may be for embryonic and fetal organs. . . . One
can even imagine without difficulty how a mother might be willing to receive into her womb as a temporary resident the embryonic clone of her desperately ill child, in order to harvest for that
child life-saving organs or tissues.

But there is always a slippery slope. Similar fears have been
raised in response to every new scientific advance. The poppy
gives us opiates to allow anesthesia during surgery but also risks
hooking addicts. Radioactivity eradicates certain cancers but
spilled in the environment can cause them. Americans worry that
terrorists might steal anthrax, smallpox, or some other biological
agent and use it to kill. Does this mean effective drugs should be
left unextracted from nature, or nature not reworked, as when we
split its atoms to generate radioactivity? Should we suspend all
microbiological research? Of course not. It simply means we must
be vigilant about the safeguards that keep potentially dangerous
science from doing harm.

Safeguards Can Work

And there is ample precedent for such safeguards. Laboratories
like my own handle highly infectious microbes within specially
designed biohazard containment facilities. We are regularly in-
spected by authorities from the university, and sometimes from
the government itself, to make sure nothing harmful is released
into the environment. Similarly, we employ types of radioactivity
that if handled improperly could contaminate the entire building.
Spot inspections occur, and labs that breach protocol risk losing
their federal funding and being shut down. Modern science does
not conform to the novels of Mary Shelley or Michael Crichton,
in which rogue researchers toil in secret rooms far from the over-
sight of universities, corporations, and the government. In fact, the
large numbers of skilled people needed for this work makes it
highly improbable that a cabal of conspirators would have access
and opportunity in the United States to conduct illegal reproductive cloning without being reported to the authorities. Does highly
improbable mean impossible? Of course not; the danger can never

be reduced to zero. But, then, the danger of reproductive cloning wouldn't be zero if we banned therapeutic cloning either. And we would forfeit a chance to save countless lives.

A Moratorium Will Not Solve the Problems

Given the arguments it deploys against therapeutic cloning, you'd think the council's majority would recommend banning it outright. But it doesn't, opting instead for a four-year moratorium. The moratorium is meant to give society time to reach a moral consensus on cloning and to give scientists time to develop other methods for developing stem cells. On both counts, however, time will not solve the problem. The report studiously avoids mentioning religion—perhaps to preempt charges that theology undergirds the anti-cloning case—but in so doing, it overstates the possibility for moral compromise. For many Americans, theology *is* central to their opposition to therapeutic cloning. Four years from now the theology of the Vatican or evangelical Protestantism is unlikely to be revised. Science will not produce data on when the soul appears, because this is a metaphysical question not amenable to experimentation; thus those who believe a cluster of cells from a manipulated egg represents sacred human life will have nothing new to consider. The council seems to anticipate new, nontheological ethical insights that will transform the cloning debate. But the report itself comprehensively delineates the secular moral positions, pro and con. It is hard to imagine new ethical insights from further debate or discussion that will turn minds one way or another, producing the "public consensus" the council's majority seeks.

The notion that a four-year moratorium will provide scientists with adequate alternative methods to develop stem cells is even more far-fetched. While research on stem cells from adults progresses, most scientists believe the therapeutic potential of these cells is not nearly so great as those derived from embryos. It is highly unlikely that adult stem cells will prove *better* than those from embryos or even as good. And most importantly, the moratorium makes it impossible to find out experimentally. As Dr. Elizabeth Blackburn, a University of California, San Francisco, biochemistry professor and a member of the council minority, writes,

It may sound tempting to impose a moratorium to get more information, since, despite very promising results, it is true, at this early stage of the research, that we still know only a little. But that information can *only* be gained by performing the same research that the moratorium proposes to halt. It has been proposed that other kinds of research will provide such answers. One cannot find out the answers about oranges by doing all the research on apples.

Studying apples may tell you something about fruit in general but certainly not about oranges specifically. Similarly, studying nonhuman embryonic stem cells from rodents or other animals during a moratorium will not produce conclusions about the scientific value of human embryonic stem cells. "Animal models," Blackburn explains, "while invaluable up to a point, cannot provide the needed information for understanding and treating a human disease." And while a moratorium prevents scientists from learning whether therapeutic cloning really can cure currently incurable diseases, Americans who suffer from those diseases will suffer in vain.

The council's moratorium is indeed a compromise—too much of one. It is a compromise of faith in our society's ability to regulate itself. It is a compromise of science's efforts to address the urgent needs of human beings with terrible maladies. It is a compromise of the future. It is a compromise of hope.

Therapeutic Cloning Should Not Be Supported

By the President's Council on Bioethics

This selection presents some of the arguments that the President's Council on Bioethics made against therapeutic cloning, arguments which eventually led to a vote for a four-year moratorium on this type of cloning. In its argument against therapeutic cloning, the council says that there are many scientific uncertainties involved in this type of cloning and that other less problematic avenues, such as nonembryonic and adult stem cell research, could replace therapeutic cloning and be equally as beneficial. In addition, the council argues that there are moral problems involved in the destruction of human embryos necessary in therapeutic cloning. The council believes that the human embryo is an organism in the process of becoming a human and therefore deserves respect and protection like any other human. Furthermore, because the embryo is helpless, the responsibility to protect it is even greater, the council claims. The Council on Bioethics was created by President George W. Bush in 2001 to investigate the ethical and policy ramifications of biomedical innovations.

Those of us who maintain—for both principled and prudential reasons—that cloning-for-biomedical-research [therapeutic cloning] *should not* be pursued begin by acknowledging that substantial human goods might be gained from this research. Although it would be wrong to speak in ways that encourage false hope in those who are ill, as if a cure were likely in the near future, we who oppose such research take seriously its potential for one day yielding substantial (and perhaps unique)

President's Council on Bioethics, *Human Cloning and Human Dignity: An Ethical Inquiry.* Washington, DC, July 2002.

medical benefits. Even apart from more distant possibilities for advances in regenerative medicine, there are more immediate possibilities for progress in basic research and for developing models to study different diseases. All of us whose lives benefit enormously from medical advances that began with basic research know how great is our collective stake in continued scientific investigations. Only for very serious reasons—to avoid moral wrongdoing, to avoid harm to society, and to avoid foolish or unnecessary risks—should progress toward increased knowledge and advances that might relieve suffering or cure disease be slowed.

We also observe, however, that the realization of these medical benefits—like all speculative research and all wagers about the future—remains uncertain. There are grounds for questioning whether the proposed benefits of cloning-for-biomedical-research will be realized. And there may be other morally unproblematic ways to achieve similar scientific results and medical benefits. For example, promising results in research with non-embryonic and adult stem cells suggest that scientists may be able to make progress in regenerative medicine without engaging in cloning-for-biomedical-research. We can move forward with other, more developed forms of human stem cell research and with animal cloning. We can explore other routes for solving the immune rejection problem or to finding valuable cellular models of human disease. Where such morally innocent alternatives exist, one could argue that the burden of persuasion lies on proponents to show not only that cloned embryo research is promising or desirable but that it is necessary to gain the sought-for medical benefits. Indeed, the Nuremberg Code of research ethics enunciates precisely this principle—that experimentation should be "such as to yield fruitful results for the good of society, unprocurable by other methods or means of study." Because of all the scientific uncertainties—and the many possible avenues of research—that burden cannot at present be met. . . .

What We Owe to the Embryo

The embryo is, and perhaps will always be, something of a puzzle to us. In its rudimentary beginnings, it is so unlike the human beings we know and live with that it hardly seems to be one of us; yet, the fact of our own embryonic origin evokes in us respect for the wonder of emerging new human life. Even in the midst of much that is puzzling and uncertain, we would not want to lose that respect or ignore what we owe to the embryo.

The cell synthesized by somatic cell nuclear transfer, no less than the fertilized egg, is a human organism in its germinal stage. It is not just a "clump of cells" but an integrated, self-developing whole, capable (if all goes well) of the continued organic development characteristic of human beings. To be sure, the embryo does not yet have, except in potential, the full range of characteristics that distinguish the human species from others, but one need not have those characteristics in evidence in order to belong to the species. And of course human beings at some other stages of development—early in life, late in life, at any stage of life if severely disabled—do not forfeit their humanity simply for want of these distinguishing characteristics. We may observe different points in the life story of any human being—a beginning filled mostly with potential, a zenith at which the organism is in full flower, a decline in which only a residue remains of what is most distinctively human. But none of these points is itself the human being. That being is, rather, an organism with a continuous history. From zygote to irreversible coma, each human life is a single personal history.

But this fact still leaves unanswered the question of whether all stages of a human being's life have equal moral standing. Might there be sound biological or moral reasons for according the early-stage embryo only *partial* human worth or even none at all? If so, should such embryos be made available or even explicitly created for research that necessarily requires their destruction—especially if very real human good might come from it? Some of us who oppose cloning-for-biomedical-research hold that efforts to assign to the embryo a merely intermediate and developing moral status—that is, more humanly significant than other human cells, but less deserving of respect and protection than a human fetus or infant—are both biologically and morally unsustainable, and that the embryo is in fact fully "one of us": a human life in process, an equal member of the species *Homo sapiens* in the embryonic stage of his or her natural development. All of us who oppose going forward with cloning-for-biomedical-research believe that it is incoherent and self-contradictory for our colleagues to claim that human embryos deserve "special respect" and to endorse nonetheless research that requires the creation, use, and destruction of these organisms, *especially when done routinely and on a large scale.*

The case for treating the early-stage embryo as simply the

moral equivalent of all other human cells is entirely unconvinc-
ing: it denies the continuous history of human individuals from
zygote to fetus to infant to child; it misunderstands the meaning
of potentiality—and, specifically, the difference between a "being-
on-the-way" (such as a developing human embryo) and a "pile of
raw materials," which has no definite potential and which might
become anything at all; and it ignores the hazardous moral precedent that the routinized creation, use, and destruction of nascent
human life would establish for other areas of scientific research
and social life.

The more serious questions are raised—about individuality, potentiality, and "special respect"—by those who assign an intermediate and developing moral status to the human embryo, and
who believe that cloned embryos can be used (and destroyed) for
biomedical research while still according them special human
worth. But the arguments for this position—both biological and
moral—are not convincing. For attempts to ground the special respect owed to a maturing embryo in certain of its developmental
features do not succeed. And the invoking of a "special respect"
owed to nascent human life seems to have little or no operative
meaning once one sees what those who take this position are willing to countenance.

We are not persuaded by the argument that fourteen days marks
a significant difference in moral status. Because the embryo's human and individual genetic identity is present from the start, nothing that happens later during the continuous development that follows—at fourteen days or any other time—is responsible for
suddenly conferring a novel human individuality or identity. The
scientific evidence suggests that the fourteen-day marker does not
represent a biological event of moral significance; rather, changes
that occur at fourteen days are merely the visibly evident culmi-
nation of more subtle changes that have taken place earlier and
that are driving the organism toward maturity. Indeed, many advocates of cloning-for-biomedical-research implicitly recognize
the arbitrariness of the fourteen-day line. The medical benefits to
be gained by conducting research beyond the fourteen-day line are
widely appreciated, and some people have already hinted that this
supposed moral and biological boundary can be moved should the
medical benefits warrant doing so.

There are also problems with the claim that its capacity for
"twinning" proves that the early embryo is not yet an individual

or that the embryo's moral status is more significant after the capacity for twinning is gone. There is the obvious rejoinder that if one locus of moral status can become two, its moral standing does not thereby diminish but rather increases. More specifically, the possibility of twinning does not rebut the individuality of the early embryo from its beginning. The fact that where "John" alone once was there are now both "John" and "Jim" does not call into question the presence of "John" at the outset. Hence, we need not doubt that even the earliest cloned embryo is an individual human organism in its germinal stage. Its capacity for twinning may simply be one of the characteristic capacities of an individual human organism at that particular stage of development, just as the capacity for crawling, walking, and running, or cooing, babbling, and speaking are capacities that are also unique to particular stages of human development. Alternatively, from a developmental science perspective, twinning may not turn out to be an intrinsic process within embryogenesis. Rather, it may be a response to a disruption of normal development from which the embryo recovers and then forms two. Twinning would thus be a testament to the resilience of self-regulation and compensatory repair within early life, not the lack of individuation in the early embryo. From this perspective, twinning is further testimony to the potency of the individual (in this case two) to fullness of form.

In Vitro Embryos Have Moral Status

We are also not persuaded by the claim that in vitro embryos (whether created through IVF or cloning) have a lesser moral status than embryos that have been implanted into a woman's uterus, because they cannot develop without further human assistance. The suggestion that extra-corporeal embryos are not yet individual human organisms-on-the-way, but rather special human cells that acquire only through implantation the potential to become individual human organisms-on-the-way, rests on a misunderstanding of the meaning and significance of potentiality. An embryo is, by definition and by its nature, potentially a fully developed human person; its potential for maturation is a characteristic it *actually* has, and from the start. The fact that embryos have been created outside their natural environment—which is to say, outside the woman's body—and are therefore limited in their ability to realize their natural capacities, does not affect either the potential or the moral status of the beings themselves. A bird forced to live in a cage its entire life may never learn to fly. But this does not mean it is less of a bird, or that it lacks the immanent potentiality to fly on feathered wings. It means only that a caged bird–like in vitro human embryo has been deprived of its proper environment. There may, of course, be good human reasons to create embryos outside their natural environments—most obviously, to aid infertile couples. But doing so does not obliterate the moral status of the embryos themselves.

As we have noted, many proponents of cloning-for-biomedical-research (and for embryo research more generally) do not deny that we owe the human embryo special moral respect. Indeed, they have wanted positively to affirm it. But we do not understand what it means to claim that one is treating cloned embryos with special respect when one decides to create them intentionally for research that necessarily leads to their destruction. This respect is allegedly demonstrated by limiting such research—and therefore limiting the numbers of embryos that may be created, used, and destroyed—to only the most serious purposes: namely, scientific investigations that hold out the potential for curing diseases or relieving suffering. But this self-limitation shows only that our purposes are steadfastly high-minded; it does not show that the *means* of pursuing these purposes are *respectful of the cloned embryos* that are necessarily violated, exploited, and destroyed in the process. To the contrary, a true respect for a being would nurture and encourage it toward its own flourishing.

It is, of course, possible to have reverence for a life that one kills. This is memorably displayed, for example, by the fisherman Santiago in Ernest Hemingway's *The Old Man and the Sea*, who wonders whether it is a sin to kill fish even if doing so would feed hungry people. But it seems difficult to claim—even in theory but especially in practice—the presence of reverence once we run a stockyard or raise calves for veal—that is, once we treat the animals we kill (as we often do) simply as resources or commodities. In a similar way, we find it difficult to imagine that biotechnology companies or scientists who routinely engaged in cloning-for-biomedical-research would evince solemn respect for human life each time a cloned embryo was used and destroyed. Things we exploit even occasionally tend to lose their special value. It seems scarcely possible to preserve a spirit of humility and solemnity while engaging in routinized (and in many cases corporately competitive) research that creates, uses, and destroys them.

The mystery that surrounds the human embryo is undeniable. But so is the fact that each human person began as an embryo, and that this embryo, once formed, had the unique potential to become a unique human person. This is the meaning of our embodied condition and the biology that describes it. If we add to this description a commitment to equal treatment—the moral principle that every human life deserves our equal respect—we begin to see how difficult it must be to suggest that a human embryo, even in its most undeveloped and germinal stage, could simply be used for the good of others and then destroyed. Justifying our intention of using (and destroying) human embryos for the purpose of biomedical research would force us either to ignore the truth of our own continuing personal histories from their beginning in embryonic life or to weaken the commitment to human equality that has been so slowly and laboriously developed in our cultural history.

The Importance of Embryos

Equal treatment of human beings does not, of course, mean identical treatment, as all parents know who have more than one child. And from one perspective, the fact that the embryo seems to amount to so little—seems to be little more than a clump of cells—invites us to suppose that its claims upon us can also not amount to much. We are, many have noted, likely to grieve the death of an embryo less than the death of a newborn child. But, then, we are also likely to grieve the death of an eighty-five-year-old father less than the death of a forty-five-year-old father. Perhaps, even, we may grieve the death of a newborn child less than the death of a twelve-year-old. We might grieve differently at the death of a healthy eighty-year-old than at the death of a severely demented eighty-year-old. Put differently, we might note how even the researcher in the laboratory may react with excitement and anticipation as cell division begins. Thus, reproductive physiologist Robert Edwards, who, together with Dr. Patrick Steptoe, helped produce Louise Brown, the first "test-tube baby," said of her: "The last time I saw her, she was just eight cells in a test-tube. She was beautiful then, and she's still beautiful now." The embryo seems to amount to little; yet it has the capacity to become what to all of us seems very much indeed. There is a trajectory to the life story of human beings, and it is inevitable—and appropriate—that our emotional responses should be different at different points in that trajectory. Nevertheless, these emotions, quite naturally and appropriately different, would be misused if we calibrated the degree of respect we owe each other on the basis of such responses. In fact, we are obligated to try to shape and form our emotional responses—and our moral sentiments—so that they are more in accord with the moral respect we owe to those whose capacities are least developed (or those whom society may have wrongly defined as "non-persons" or "nonentities").

In short, how we respond to the weakest among us, to those who are nowhere near the zenith of human flourishing, says much about our willingness to envision the boundaries of humanity expansively and inclusively. It challenges—in the face of what we can know and what we cannot know about the human embryo—the depth of our commitment to equality. If from one perspective the fact that the embryo seems to amount to little may invite a weakening of our respect, from another perspective its seeming insignificance should awaken in us a sense of shared humanity. This was once our own condition. From origins that seem so little came our kin, our friends, our fellow citizens, and all human beings, whether known to us or not. In fact, precisely because the embryo seems to amount to so little, our responsibility to respect and protect its life correspondingly increases. As [philosopher] Hans Jonas once remarked, a true humanism would recognize "the inflexible principle that utter helplessness demands utter protection."

Reproductive Cloning Is Beneficial

By Panayiotis Zavos

In the following selection, originally given as testimony before the Subcommittee on Criminal Justice, Drug Policy and Human Resource House Government Reform on May 15, 2002, Dr. Panayiotis Zavos, a reproductive specialist, presents arguments for human reproductive cloning. Zavos says that human reproductive cloning would allow infertile couples to have their own biological children, which is a right guaranteed to every American. He believes a lack of understanding about cloning has created an unreasonable fear of the technology and claims that some scientists in the field have misled the public for their own gain. In addition, Zavos says that laws in the United States will not stop human cloning; American couples will just travel to other nations where cloning is legal. He argues that, like the initial banning of in vitro fertilization, a ban on human cloning would eventually be lifted as scientists demonstrate the safety and effectiveness of the technology. Zavos stresses that the United States should take the lead in this important new technological arena.

Dr. Panayiotis Zavos has been involved in the development of several new technologies in the animal and human reproductive area. He is the head of the Zavos Diagnostic Laboratories and director of the Andrology Institute of America. Zavos is also associate director of the Kentucky Center for Reproductive Medicine and IVF and professor emeritus of Reproductive Physiology and Andrology at the University of Kentucky.

Panayiotis Zavos, testimony before the Subcommittee on Criminal Justice, Drug Policy, and Human Resources, Washington, DC, May 15, 2002.

I am a reproductive specialist and scientist that has dedicated the last 24 years of my life in helping infertile couples have children and complete their biological cycle. In January 2001, we have announced the possibility of using reproductive regeneration technologies as a means of treating infertility, and our intention to develop these technologies in a safe and responsible manner. However, we have received great opposition from fellow scientists, news media and the general public. It seems that the great opposition is due to the lack of complete understanding and comprehension of what in actuality human cloning really is all about. The British Medical Association however, has so appropriately stated: "Public hostility to human reproductive cloning may be based on an illogical transient fear of a new technology." Much of the confusion is caused by the variance in opinions coming from different scientific sources, politicians, news media, and Hollywood. Due to the limited knowledge of these technological and medical procedures in the scientific community, we have organized, hosted and attended meetings involving scientists from all over the world to discuss and debate the issues of human reproductive regeneration. We have even presented our intentions before the Congress of the United States. . . .

Infertility affects approximately 10–15% of couples of reproductive age throughout the developing world. Assisted reproductive technologies (ART) have played a major role in treating various causes of infertility. In fact, about 65% of the couples who seek medical help will eventually succeed in having a child. However, in cases where there are no sperm or eggs present (possibly due to loss of testicular or ovarian function), the only options these couples face are sperm donation, oocyte donation or adoption. These are difficult choices for couples to make and many do not want to use sperm or egg sources other than their own or do not wish to consider adoption. Reproductive regeneration (RR), which is synonymous to reproductive cloning, can therefore play a very real role in the treatment of severe male or female infertility in couples that wish to have their own biological children.

After a lot of time, money and suffering, many of the infertile couples have been able to have children using present IVF [in vitro fertilization] techniques. Personally, it has given me great satisfaction to assist them in the creation of their own families. However, some of these infertile couples have not been able to experience the joy of creating their own families because the present technologies are not advanced enough to help them. For them, human reproductive cloning is the only way they can have their own children. As a reproductive specialist and a scientist who cares about their plight, I am trying to develop safe techniques of human cloning so they can have the healthy babies they want. Mr. Chairman, am I wrong in wanting to help couples become parents?

If you care about these unfortunate infertile couples, why are you considering legislation that would make both them and the people that are trying to help them, criminals? Criminalizing human reproductive cloning in the United States will only make it less safe and more costly for these infertile couples. They will be forced to travel outside the United States to pursue their dream of creating a family. After all, according to the Americans with Disabilities Act (ADA), infertility is a disability and reproduction is a major life activity for the purposes of the ADA. In light of this, it is the right of each and every American citizen to bear a child.

Cloning Cannot Be Curbed

Mr. Chairman, experts state repeatedly that history proves the point very clearly that scientists will clone even if [President George W.] Bush and the Congress forbid it. The House of Representatives may vote against human cloning but that will not stop scientists from doing it and people from wanting it. The American Society for Reproductive Medicine (ASRM) of which I am a long standing member, recently stated that "thousands of years of human experience have shown us that governments cannot bottle up human progress, even when you want to" and that "there is every reason to believe that if passed, this kind of prohibition would not be effective". In another case made by a infertility patient, who wants her own genetic baby so badly that she would go wherever she had to, in order to clone either herself or her husband "if they called me right now and said, 'We're paying for everything and giving you the chance to have your own genetic child,' I would be on a plane so fast it's not even funny," she said. In the words of a bioethicist "The best way to control this research is to fund it by the federal government, because then you create rules," and in my words, Mr. Chairman, this Genie is out of the bottle and it keeps getting bigger by the hour. There is no way that this Genie is going back into the bottle. Let us find ways to develop it properly and disseminate it safely.

Banning human reproductive cloning in the United States will not stop human cloning. In fact, the first cloned pregnancy may have occurred already. If you institute a ban, all that will happen is exactly what happened when the first IVF baby was born in 1978. The United States banned IVF when it first came out and then after several years, decided it had made a mistake and spent the next several years catching up with the technology that was advanced in other countries. The only people that suffered were the infertile U.S. couples who were unable to have children or had to travel outside of the United States to receive these treatments. Let us show the proper compassion for those suffering American infertile couples. Let us give them some hope and let us not turn our backs on them. They deserve something better than that.

If you are concerned about the risks of human cloning, the proper approach is to fund it and then institute regulations that will insure that human cloning is done properly with a minimum of risk for the baby just as is done in other medical or drug innovations. This is what our team is working on and we will not go forward with human cloning until the risks are comparable with other IVF procedures. Of course, because of the present political climate in the United States, we have been forced to look elsewhere in the world for a proper venue. We have no intentions of doing this in the USA whether any legislation is passed for or against this technology. Furthermore, Mr. Chairman, we have no intentions of breaking the laws of this country or any other country to accomplish this. We are law abiding citizens of this great Nation of ours, but we are a compassionate group of people that wish to help our fellow man and woman have the gift of life. The gift of life that most of us have been so fortunate to have, enjoy and take for granted. Let us not be so uncompassionate and so insensitive to tell those people that we are not willing to listen to them and unwilling to help them. This is not what our country's constitution and principles are based on. We believe in creating families, not preventing them. In God we trust!

Reproductive Regeneration as a Means of Infertility Treatment

The incidence of developmental abnormalities following natural sexual reproduction in humans is 3% and is significantly higher when maternal age is over 40. As recently reported in the *New En-*

gland Journal of Medicine, the risks are even greater from IVF and other more advanced ART procedures yielding more than 30,000 children per year in the USA. It is vividly clear that thousands of potential parents accept these risks to conceive a child. If human reproductive regeneration is banned as a reproductive technique on safety grounds, then we may find ourselves in the untenable position of having banned all reproductive techniques which suffer equal or higher risks, thereby, possibly even banning natural sexual reproduction with its 3% risk, a situation that the majority of people would consider ridiculous. It appears reasonable to suggest that the incidence of developmental abnormalities as to the safety of human reproductive regeneration is negligible when compared to current risks associated with IVF and other ART procedures.

It is quite evident to us along with other competent human reproductive specialists that with further elucidation of the molecular mechanisms involved during the processes of embryogenesis, careful tailoring of subsequently developed culture conditions and manipulation strategies, and appropriate screening methods, will eventually allow infertile couples to safely have healthy, genetically related children through SCNT [somatic cell nuclear transplant] methods.

The Opponents of Human Cloning or Reproductive Regeneration

The most prominent opponents to human reproductive regeneration and spokesmen for animal cloning are Drs. Ian Wilmut from the Roslin Institute and Rudolph Jaenisch from the Massachusetts Institute of Technology (MIT), who have misled and have misdirected the public and its leadership for their very own gains, whatever those gains might be. They have repeatedly stated that the application of animal cloning technologies to humans, is extremely dangerous, not because of ethical and social implications, but because of the foreseeable possibility that cloning humans might result in a very high incidence of developmental abnormalities, large offspring syndrome (LOS), placental malfunctions, respiratory distress and circulatory problems, the most common causes of neonatal death in animals. They also noted that the rate of success as an ART method is extremely low, being only 3%. Furthermore, they state that because since the production of Dolly the sheep in

1995, they have not improved on these technologies themselves, they have concluded that reproductive regeneration is not safe and efficient for use in humans, and would like for the world to believe this. Let us examine the facts as they appear.

If one reviews the animal cloning literature, one can deduce that the poor cloning success rates noted by the "animal cloners" are mainly due to experiments that were poorly designed, poorly executed, poorly approached, and poorly understood and interpreted. These experiments were mostly done under non-sterile and uncontrolled environments and having a "hit-and-miss" type of outcome. Also, when the cloned animals died, no clear view of their cause of death was ascertained. In short, their experimentation methods lacked the seriousness of purpose that is vital when performing similar studies in humans. Furthermore, the same scientists responsible for Dolly, the sheep, now plan to utilize similar crude technologies to experiment on cloned human embryos for medical purposes.

According to a recent article in *Time* magazine, Wilmut and Jaenisch stated "animal cloning is inefficient and is likely to remain so for the foreseeable future". On the contrary, a number of studies have already demonstrated far higher rates of success and, in some cases, matching or exceeding successes noted in human IVF today. Also, if history is any indicator, one can reasonably expect that further refinements to the cloning process will improve efficiency rates. Scientists have reported success rates of 32% in goats and 80% in cows since 1998, as opposed to the poor 3% success rate Wilmut obtained when cloning Dolly in 1995. Furthermore, scientists at Advanced Cell Technologies in Worcester, Massachusetts, in association with others, have recently produced 24 cloned cows, that were all normal and healthy and have survived to adulthood. Despite the overwhelming data that exists showing refinements in the RR technology that yield improving success rates, Wilmut and Jaenisch still insist that it is inefficient based upon their poor success using very crude and uncontrolled experimental techniques, almost seven years ago. One can only but question their motives for their illogical arguments. They do not seem interested in developing and refining techniques, but they rather seem to have immense private interests and want to patent and control the technologies for themselves. Interestingly enough, the Roslin Institute scientists who cloned Dolly the sheep have changed their agenda on the cloning subject and have stated recently that they plan to seek permission to experiment on cloned human embryos for medical purposes. What are their true motives?

Animal Cloning vs. Human Reproductive Regeneration

It has been very clearly shown that animal cloning and its difficulties appear to be species-specific, and the data cannot be extrapolated with a great degree of accuracy to the human species. In a recent study by scientists from Duke University Medical Center, it was demonstrated that it may be technically easier and safer to perform somatic cell nuclear transfer (SCNT) in humans than in sheep, cows, pigs and mice because humans possess a genetic benefit that prevents fetal overgrowth, one of the major obstacles encountered in cloning animals.

The genetic benefit is based on the fact that humans and other primates possess two activated copies of a gene called insulin like growth factor II receptor (IGF2R). Offspring receive one functional copy from each parent as expected. However sheep, pigs, mice and virtually all non-primate mammals receive only one functional copy of this gene because of a rare phenomenon known as genomic imprinting in which the gene is literally stamped with [a] marking that turns off its function. Since humans are not imprinted at IGF2R, then fetal overgrowth would no be predicted to occur if humans were cloned. If this theory is correct, the incidence of developmental abnormalities following human SCNT would be significantly lower. Also, the authors concluded that the data showed that one does not necessarily have these problems in humans. This is the first concrete genetic data showing that the cloning process could be less complicated in humans than in sheep.

The Political Status on Cloning

In the United States, the House passed in July, 2001 the Weldon Bill or the Human Cloning Prohibition Act of 2001 (bill H.R. 2505). This bill would prohibit any person or entity, in or affecting interstate commerce, from performing or attempting to perform human cloning, participating in such an attempt, shipping or receiving the product of human cloning, or importing such a product. The bill currently pending in the US Senate, S 790, written by Sen. Sam Brownback (R-Kansas), would criminalize all cloning with a fine of up to $1 million and 10 years in prison and it is almost identical to the bill (H.R. 2505) passed by the House in July 2001. The Council of Europe has introduced a protocol that prevents any abuses of such techniques by applying them to humans, banning "any intervention seeking to create a human being genetically identical to another human being, whether living or dead". Finally, the Protocol leaves it to countries' domestic law to define the scope of the term "human being". In April 24, 2001, England has banned "reproductive regeneration" but not "therapeutic cloning".

The political situation with cloning in general remains very fluid, mainly because of the inability of the politicians to understand, comprehend and act decisively on the issues that cloning presents to society. After all, their inability to act decisively may have a great deal to do with their resistance to debate and face the facts that humans will be cloned.

In his speech to the American public, President Bush made an appeal for a global ban on cloning, whether it be for therapeutic or reproductive cloning, on the basis that we should not use people for "spare parts" and we should not "manufacture people". Reproductive cloning does neither. As opposed to therapeutic cloning which results in the inevitable death of an embryo once the stem cells have been removed, reproductive cloning aims to protect and preserve life in allowing the embryo to grow and be implanted into the uterus for a subsequent pregnancy. From an ethical point of view, there is no destruction of life.

Quoting President Bush: "Life is a creation, not a commodity. Our children are gifts to be loved and protected, not products to be designed and manufactured. Allowing cloning would be taking a significant step toward a society in which human beings are grown for spare body parts, and children are engineered to custom specifications; and that's not acceptable". And that's not acceptable to us either, Mr. Chairman! We agree with President Bush and uphold the sanctity of human life. Reproductive cloning does not involve the destruction of human embryos, nor does it modify or "engineer" the genetic code to custom specifications. Reproductive cloning involves employment of similar technology used for Intra Cytoplasmic Sperm Injection (ICSI), which is routinely employed in IVF centers throughout the world. The only difference is that instead of using a sperm cell from the father, scientists can use a somatic cell nucleus and inject it into the mother's anucleated egg. The resulting embryo would have its genetic makeup from the father, but the expression of the genetic code and characteristics and personality of the baby born will be completely different and unique. Reproductive cloning is nothing more than another modality for the treatment of human infertility in giving the gift of life to a childless couple that have exhausted all other choices for having a child. What is so wrong about this?

Is History Repeating Itself?

This is not the first time that the scientific community has had to deal with controversial issues regarding new technologies. Exactly the same events happened with IVF in the Kennedy Institute in Washington in 1978. Professor Robert Edwards and Dr. Patrick Steptoe were faced with such criticism from hundreds of reporters, senators, judges, scientists and doctors, when they proposed the idea of in vitro fertilization. The language and accusations were the same as what we face today, Including "they ignored the sanctity of life, performed immoral experiments on the unborn," "subject to absolute moral prohibition," "no certainty that the baby won't be born without defect" and to "accept the necessity of infanticide. There are going to be a lot of mistakes".

Twenty-four years later, the exact opposite of everything the "experts" predicted happened. IVF has become an acceptable and routine treatment of infertility worldwide. The abnormalities that were expected to have been unacceptable proved to be the same, if not less than with natural conception. Ironically, those critics of IVF have become the "pioneers" of IVF. These same critics might have delayed the introduction of IVF but their actions mostly harmed patients, and also the medical and scientific community. I am certain that the reproductive cloning procedures will follow in the same footsteps. Recently, I have had the opportunity to openly debate Professor Robert Winston from the UK, on the issue of human reproductive cloning at an Oxford Union Debate at Oxford University. Ironically enough, he was one of the leaders originally opposed to IVF, and who is currently a leading IVF specialist in Britain. The technology that he was vehemently opposed to, almost twenty-five years ago, is now the very same technology that he uses to earn a living. Once reproductive regeneration is commonplace in the ART treatment market, will he, along with all the other critics, "jump" on the bandwagon and offer this new technology in their own IVF centers? I believe so. They have done it before and they can do it again. Mr. Chairman, we can not afford to behave this way and most importantly wish to repeat the same mistake.

Cloning Will Come to Be

As Professor Robert Edwards, the great English scientist who helped create the world's first test-tube baby in 1978, so eloquently prophesied recently "Cloning, too, will probably come to be accepted as a reproductive tool if it is carefully controlled". No doubt, humans will be produced via reproductive regeneration. Recent scientific and technological progress demonstrates that very clearly. Similar to IVF, the technology of reproductive regeneration will advance, techniques will be improved, and knowledge will be gained. Reproductive regeneration's difficult questions can be answered only through a dedicated pursuit of knowledge and an exercise of our willful rationality, and in the end, the answer to the debate over human nature may be simply that man's nature is the product of his own will.

Mr. Chairman, science has been very good to us and we should not abandon it now. Consider why America has the best medical care in the world. It is because we have the freedom to investigate, research and market the latest medical techniques, all within proper procedures and safeguards. This is not the time to panic and try to turn back the clock. The Genie is already out of the bottle. Let's make sure it works for us, not against us. Let's do it here. Let's do it right.

By banning cloning, America will be showing the world that she is hesitant and/or reluctant to take the lead in this new arena of technological advancement. The world today is looking at the most powerful nation on Earth for leadership on this issue, and walking away from it by banning it is not a sign of leadership, but cowardice. Do not let the future of this technology slip away through our fingers, because we are too afraid to embrace it. I believe that it is the right of the American people to choose whether or not they want to have this technology available to them. Let us educate ourselves and debate the issues and not make irrational decisions based upon fear of a new technology. Banning this technology would only give our enemies license to use it to their advantage. Let us learn from history and forge ahead in this brave new world as leaders, not spectators, the American way.

The Dangers of Reproductive Cloning

By the President's Council on Bioethics

The President's Council on Bioethics was created by President George W. Bush in 2001 to address the ethical and policy ramifications of biomedical innovations. In the following selection, taken from a working paper, the council staff presents some of the many arguments against reproductive cloning. The staff suggests that reproductive cloning is dangerous in light of the many problems and failures scientists have had in cloning animals. For example, many cloned animals die prematurely. The council also claims that cloning could lead to eugenics, the practice of altering (with the aim of improving) the genetic makeup of future generations. This practice, the staff believes, would blur the sense of what is and is not human. In addition, reproductive cloning, which tampers with the very complex system of natural human reproduction, may result in profound alterations of human nature and may diminish the diversity of the human gene pool. Reproductive cloning could also transform human procreation into a human manufacturing process. Because scientists or parents could design children with specific characteristics, children would become products of a dehumanizing manufacturing process, and society's attitudes toward them would reflect the childrens' subservient status. Furthermore, because reproductive cloning affects more than the individuals involved, the council believes society must address the issue for the well-being of future generations.

President's Council on Bioethics, *Arguments Against Reproductive Cloning*. Washington, DC, January 2002.

Opposition to human "reproductive" cloning falls into several general categories, some of which address the more familiar questions of safety, consent, and individual rights, but most of which focus mainly on the larger human questions at stake in making this decision.

Safety and Health of Children and Mothers

The first of these is a concern raised by nearly everyone on all sides of the cloning debate: the safety of all involved. Even most proponents of "reproductive" cloning generally qualify their support with a caveat about the safety of the procedure. Almost no one argues that cloning is presently safe enough to attempt on human beings, and the example of cloning experiments in other mammals strongly suggests that human "reproductive" cloning is, at least for now, far too risky to attempt. Safety concerns revolve around potential dangers to the cloned child, as well as to the egg donor and the surrogate mother.

Risks to the cloned child must be taken especially seriously, not least because—unlike the risks to the egg donor and surrogate mother—they cannot be accepted knowingly and freely by the person who will bear them. The risks to the cloned child have at this point led nearly everyone involved in the debate to consider cloning thoroughly unsafe. In animal experiments to date, only approximately 5 percent of attempts to clone have resulted in live births, and a substantial portion of those live-born clones have suffered complications that proved fatal fairly quickly. Longer term consequences are of course not known, since the oldest successfully cloned mammal is only approximately five years of age. Some medium-term consequences, including premature aging, immune system failures, and sudden unexplained deaths, have already become apparent in some cloned mammals.

Furthermore, there are concerns that a cell from an individual who has lived for some years may have accumulated genetic mutations which—if used in the cloning of a new human life—may predispose the new individual to certain sorts of cancer and other diseases.

Along with these threats to the health and well-being of the cloned child, there appear to exist some risks to the health of the egg donor (particularly risks to her future reproductive health caused by the hormonal treatments required for egg donation) and risks to the health of the surrogate mother (for instance, animal experiments suggest a higher than average likelihood of over-weight offspring, which can adversely affect the health of the birth-mother).

These concerns have convinced most of those involved in the field that attempts at "reproductive" cloning at present would constitute unethical experimentation on human subjects, and should be forbidden. These considerations of safety were among the primary factors that led the National Bioethics Advisory Commission to call for a prohibition of human "reproductive" cloning in 1997, and the evidence in support of such concerns has only grown since then.

Nonetheless, as the NBAC report also articulated, safety concerns may well be temporary, and could prove amenable to technical solutions. These concerns constitute a persuasive argument for a moratorium or temporary ban; but in themselves they do not get beyond the technicalities of cloning to the deep moral, social, and ethical issues involved. The present Council, given its membership and its charter, may be more inclined to consider the meaning of cloning beyond its technical feasibility and its safety record, and to look at the more permanent significance of cloning as an activity that reflects on those who engage in it and on the society that permits or encourages it. It is in those areas of inquiry that serious and permanent objections to cloning arise with the greatest force.

The Violation of Individual Rights

Beyond physical safety, the prospect of "reproductive" cloning also raises concerns about a potential violation of the rights of individuals, particularly through a denial of the right to consent to the use of one's body in experimentation or medical procedures.

Consent from the human clone itself is of course impossible to obtain. It may be argued, on the one hand, that no one consents to his own birth, so concerns about consent are misplaced when applied to the unborn. But as [health professor] George Annas and others have argued, the issue is not so simple. For reasons having much to do with the safety concerns raised above and the social and psychological concerns to be addressed below, an attempt to clone a human being would expose the cloned individual-to-be to great risks of harm, in addition to, and different from, those accompanying other sorts of reproduction. Given the risks, and the fact that consent cannot be obtained, the ethical choice may be to avoid the experiment.

Against this point it might be said that the alternative to cloning is for the cloned individual not to exist at all, and that no one would prefer non-existence to the chance at life. Such an argument, however, could easily come to be used as an excuse for absolutely any use and abuse of embryonic or newborn life. Giving life to an individual does not grant one the right to harm that individual. It is true that the scientist cannot ask an unconceived child for permission, but this puts a burden on the scientist, not on the child. All that the scientist can know is that he or she is putting a newly created life at enormous risk; and given that knowledge, the ethics of human experimentation suggest that the best option is to avoid the procedure altogether.

Indeed, an inquiry into the purpose and meaning of consent may well support this point. Why, after all, does society insist upon consent as an essential principle of the ethics of scientific research? The requirement for consent is not quite an end in itself. It exists to protect the weak and the vulnerable, and particularly to protect them from the powerful. It would therefore be morally questionable, at the very least, to choose to impose potentially grave harm on an individual, even by the very act of choosing to give them life.

A separate question of consent arises in light of the possibility that individuals, living or dead, may be used as sources of DNA for a cloning procedure without their permission or even their knowledge. Unlike other forms of reproduction, including assisted reproduction and in vitro fertilization, cloning could be carried out with DNA from individuals who have chosen to be involved in a reproductive procedure. While an egg or sperm donor may not consent specifically to have the egg or sperm used in a particular reproductive procedure, the donor has knowingly donated the sperm or egg, and thus has consented to having them used in such procedures in general. But since, at least in theory, the nucleus of any cell from a given individual could be inserted into an egg to clone that individual, and since cells can be obtained illicitly with relative ease, individuals may find themselves cloned without their knowledge or approval, and thus essentially forced to reproduce without their consent.

A subset of this problem involves the cloning of the deceased, who, like the unborn, cannot provide or deny consent. Among the possible uses for cloning suggested by proponents of human "reproductive" cloning is the opportunity it offers for parents of a deceased child, or the family of any deceased individual to clone that individual. Such an action, taken without prior permission, could be a patent violation of the principles of reproductive consent, though of course the individual whose rights are violated would be unable to object.

In these different ways, the long-standing insistence on obtaining consent for medical procedures, and particularly reproductive ones, could be seriously undermined by the advent of human "reproductive" cloning.

Eugenics and Enhancement

Human "reproductive" cloning could also come to be used for eugenic purposes: that is, in an attempt to alter (with the aim of improving) the genetic constitution of future generations. Indeed, that is the stated purpose of some proponents of "reproductive" cloning, and has been at the heart of much support for the concept of "reproductive" cloning for decades. Proponents of eugenics were once far more open regarding their intentions and their hopes to escape the uncertain lottery of sex and reach an era of controlled and humanly directed reproduction, which would allow future generations to suffer fewer genetic defects and to enjoy more advantageous genotypes. In the present debate, the case for eugenics is not made quite so openly, but it nonetheless remains an important driving motivation for some proponents of human cloning, and a potential use of "reproductive" cloning.

Cloning can serve the ends of eugenics either by avoiding the genetic defects that may arise when human reproduction is left to natural chance; or by preserving and perpetuating outstanding genetic traits. In the future, if techniques for precise genetic engineering become available, cloning could be useful for perpetuating the enhanced traits created by such techniques, and for keeping the "superior" man-made genotype free of the flaws that sexual reproduction might otherwise introduce.

The darkest side of eugenics is of course familiar to any student of the twentieth century. Its central place in Nazi ideology, and its brutal and inhuman application by the Third Reich, have put that science largely out of favor. No argument in today's cloning debate bears any resemblance to those of Hitler's doctors. But by the same token, it is not primarily the Nazi analogy that should lead us to reject eugenics.

It is a less dark side of eugenic science that threatens to confront us. This side is well-intentioned but could prove at least as dangerous to our humanity. The eugenic goal of "better" and "healthy" children combined with modern genetic techniques threatens to blur and ultimately eliminate the line between therapy and enhancement. Medicine is guided by the natural standard of health. It is by this standard that we judge who is in need of medical treatment, and what sort of treatment might be most appropriate. The doctor's purpose is to restore a sick patient to health. Indeed, we even practice a kind of "negative" eugenics guided by this standard: as when parents choose to abort a fetus who has been diagnosed with a serious genetic disease. This "negative eugenics" may be morally problematic in itself, but it is at least a practice that is informed by a standard of health.

The "positive" eugenics that could receive a great boost from human "reproductive" cloning does not seek to restore human beings to natural health when they are ill. Instead, it seeks to alter humanity, based upon a standard of man's (or some men's) own making. Once the natural goal of health has been blurred out of existence, medicine will come to serve only ends designed by human will, and thus may have no limits, may feel no constraints, and may respect no barriers. Reproduction itself might come to serve one or another purely man-made end, and future generations may come to be products of our artful and rational design more than extensions of our humanity. All of this may well be guided by what plainly seem like good intentions: to improve the next generation, to enhance the quality of life of our descendants, to let our children do more than we ourselves could do. But in the process, we stand to lose the very means by which to judge the goodness or the wisdom of the particular aims proposed by a positive eugenics. We stand to lose the sense of what is and is not human; a set of limits on our hubris; a standard against which to judge the legitimacy of certain human actions. All of these, along with the specific traits and characteristics done away with in the process of eugenic enhancement, could be lost. "Reproductive" cloning may well contribute to these losses.

Eugenics, and cloning itself, may also contribute to an unhealthy belief in genetic determinism, which could have profoundly negative social consequences. As we become better able to manipulate and to control human genotypes, we may tend to place greater importance on the genotype, wishing to secure for our descendants every possible advantage. To a man with a hammer, everything begins to look like a nail. To a society armed with the power to control and change the genome, the genome will suddenly look very much in need of control and change. The ability to manipulate the genotype of an individual may tend to convince us of the supreme importance of genetics in shaping an individual, which in turn may lead us to want more control, in a self-intensifying cycle pointing toward an increasing surrender to an ideology of genetic determinism. Whether or not our control of the genome actually turns out to give us much control over individuals, our new ability may of itself be enough to lead us to place undue importance on genetics.

It is essential to realize that many of the social concerns raised below could result from this very attitude. An excessive focus on the importance of the genotype would exacerbate the social and psychological pressures to which a cloned individual may be subject.

In addition, eugenics may also open the road to a new inequality, by which only those who can afford it can procure advantages for themselves and their descendants into future generations. A situation in which only the rich can grant their children high IQs, broad shoulders and long lives would prove unbearable to a liberal democracy, and might either lead to serious social tension or more likely to a government entitlement to genetic enhancement and manipulation—managed by the state.

By serving the ends of eugenics, "reproductive" cloning may open the door to all of these various difficulties.

Respect for Nature

Cloning also raises a number of concerns about humanity's relation with the natural world. The precautionary principle, which informs the ideals of the environmental movement, may have something to say to us about cloning. It urges us to beware of the unintended consequences of applications of human power and will—particularly over nature. Natural systems of great complexity do not respond well to blunt human intervention, and one can hardly think of a more complex system than that responsible for human reproduction. This principle suggests that geneticists should not pretend to understand the consequences of their profound alterations of human nature, and lacking such understanding they should not take actions so drastic as the cloning of a human child.

The ethic of environmentalism also preaches a respect for nature as we find it, and argues that the complex structure of the natural world has much to teach us. Such an ethic therefore disapproves of efforts aimed at simply overcoming nature as we find it, and imposing a man-made process over a slowly evolved natural process. It opposes the hubristic overconfidence inherent in the cloning project, and fears that such a project may erase the boundary between the natural and the technological.

In addition, cloning, in the unlikely event it should become commonplace, may diminish the diversity of the human gene-pool. Sexual reproduction introduces unique combinations of genes into the human gene-pool, while eugenic cloning aimed at reproducing particular genotypes will tend to diminish that diversity, and with it the "strength" of the species. Eugenic enhancement may thus "weaken" future generations.

Manufacturing Humans

"Reproductive" cloning could also represent an enormous step in the direction of transforming human procreation into human manufacture.

In natural procreation, two individuals come together to give life to a new individual as a consequence of their own being and their own connection with one another, rather than merely of their will. They do not design the final product, they give rise to the child of their embodied selves, and they therefore do not exert control over the process or the resulting child. They beget something that is in essence like themselves; they do not make something that is in essence their own. The product of this process, therefore, stands beside them fully as a fellow human being, and not beneath them as a thing made by them with only their own purposes in mind. A manufactured thing can never stand beside its human maker as an equal, but a begotten child does stand equally beside its parents. The natural procreative process allows human beings—through the union of male and female—to make way for fellow human beings, to whom they give rise, but whom they do not make. It thus endows each new generation with the dignity and freedom enjoyed by all that came before it.

Even most present forms of partially artificial reproduction, including IVF [in vitro fertilization], essentially imitate this natural process, and while they do begin to introduce the characteristics of manufacture and industrial technique, they cannot claim to control the final outcome as an artisan might shape his artifact. The end they serve is still the same—the birth of a child from the sexual union of seed from two progenitors. Reproduction with the aid of such techniques therefore still at least implicitly arises from (and gives rise to) a willingness to accept the product of a process we do not control. In this sort of procreation, children emerge out of the same mysterious process from which their parents came, and therefore are not mere creatures of their parents.

Human "reproductive" cloning, and the forms of human manufacture it might make possible, could be quite different. Here, the process would begin with a very specific end-product in mind, and would be tailored to produce that product. Scientists or parents would set out to produce specific individuals for particular reasons, and the individuals might well come to be subjected to those reasons. The procreative process could come to be seen as a means of meeting some very specific ends, and the resulting children would be products of a designed manufacturing process: means to the satisfaction of a particular desire, or to some other end. They would be means, not ends in themselves.

Things made by man stand subservient to the man who made them. Manufactured goods are always understood to have been made to serve a purpose, not to exist independently and freely. Scientists who clone (or even merely breed) animals make no secret of the instrumental purposes behind their actions—they act with specific instrumental ends in mind, and the resulting animals are means to that preexisting end. Human cloning threatens to introduce the same approach and the same attitudes into human procreation.

The transformation of human procreation into human manufacture could thus result in a radical dehumanization of the resulting children, as well as of those who set out to clone, and by its effect on societal attitudes also a dehumanization of everyone else. When we become able to look upon some human beings as manufactured goods, no matter how perfect, we may become less able to look upon any human beings as fully independent persons, endowed with liberty and deserving of respect and dignity. . . .

Identity and Individuality

The above-stated concerns about the consequences of cloning as a manufacturing process lead into broader and more serious concerns about the mental and emotional life and the personal and social relations of the individual produced by a "reproductive" cloning procedure. These concerns would apply even if cloning was only conducted on a small scale.

The natural procreative process is uniquely capable of endowing new human beings with a combination of rootedness and family bonds on the one hand, and independence and individuality on the other. Our genetic uniqueness and our genetic relatedness to others both mirror and ground this social human truth: Each of us has a unique, never-before-enacted life to live with a unique trajectory from birth to death; and each of us owes our existence and our rearing to those who have come before and who have brought us into being and taken responsibility for our existence and our rearing. By nature, every child is tied to two biological parents, and that child's unique genetic identity is determined by what is essentially a chance combination of these parents' genotypes. Each child is thus related equally and by the closest of natural bonds to two adult human beings and yet each child is genetically unique. Both these characteristics, and the procreative nature of humanity from which they arise and to which they point, help give shape to the psyche of each of us, and to the human institutions that allow us to thrive.

Our genetic uniqueness, manifested externally in our looks and our fingerprints and internally in our immune systems, is one source of our sense of freedom and independence. It symbolizes our autonomy and it endows us with a sense of possibility. Each of us knows that no one has ever had our unique combination of natural characteristics before. We know that no one knows all the potentialities contained within that combination. A cloned child, however, will live out a life shaped by a genotype that has already lived. However much or little this may actually mean in terms of hard scientific fact, it could mean a great deal to that individual's experience of life. He or she may be constantly held up to the model of the source of his or her cloned genotype, or may (consciously and unconsciously) hold himself or herself up to that model. He or she would be denied the opportunity to live a life that in all respects has never been lived before, and (perhaps more importantly) might know things about his or her own genetic destiny that may constrain his or her range of options and sense of freedom. . . .

The Impact on Society

Cloning is a human activity, which affects not only those who are cloned or who are clones, but also the entire society that allows or that supports (and therefore that engages in) such activity—as would be the case with a society that allows some of its members to practice slavery, to take a most extreme example. The question before us is whether "reproductive" cloning is an activity that we, as a society, should engage in. In addressing this question, we must reach well beyond the rights of individuals, and the difficulties or benefits that cloned children or their families might encounter. The question we must face has to do with what we, as a society, will permit ourselves to do. When we say that "reproductive" cloning may erode our respect for the dignity of human beings, we must say that we, as a society that engages in cloning, would be responsible for that erosion. When we argue that vital social institutions could be harmed, we must acknowledge that it is we, as a society that clones, that would be harming them. We should not ask if "reproductive" cloning is something that some people somewhere should be permitted to do. We must ask if cloning is something that all of us together should want to do or should allow ourselves to do. Insofar as we permit cloning in our society, we are the cloners and the cloned, just as we are the society affected by the process. Only when we see that do we understand our responsibility in crafting a public policy regarding human "reproductive" cloning.

Since we are the ones acting to clone, we must further realize that our actions will affect us not only in what they directly do to us, but also in the way they shape our thinking. A society that clones human beings is a society that thinks about human beings differently than a society that refuses to do so. We must therefore also ask ourselves how we as a society prefer to think of human beings.

These sorts of questions are not easy for a modern liberal polity to contend with. We are not accustomed to thinking in these terms, and we are not comfortable using them in political discourse. But cloning (along with the accompanying broader issues raised by contemporary biotechnology) forces us, as few other matters do, to think this way, because cloning is a human activity that threatens (or promises) to affect the very nexus of human societies: the junction of human generations. Liberal society will not and should not seek to regulate every sort of human activity, but it cannot help but involve itself in those that directly affect its highest and most urgent tasks: tasks like its own perpetuation and the transmission of its ideals and way of life to future generations. Doing nothing about such a subject is not an option. If we as a society refrain from considering the question entirely, we would—implicitly— be saying yes to cloning, with all that such a statement would entail. We face the choice only of engaging in cloning or forbidding it, and the option we select will say a lot about us. Given the issues involved, there is no neutral ground for the polity to hold in this particular debate.

Society exists beyond individuals, beyond generations. And among the highest tasks of any society is the management of its relation to the future, and the transmission of its institutions and its ideals to the next generation. As we have seen, it is here, at this vital junction of the generations, that cloning threatens to wreak havoc, and this junction is insufficiently protected by the market and by private interests. It is here that cloning poses a special challenge to society, and it is here that politics becomes important in meeting that challenge.

Politics becomes important because politics is the means that a free society has at its disposal to protect its common interests, to serve its common needs and to express its common will. There is, in modern free societies, a very reasonable reticence to bring politics into what is usually quite properly thought of as the very private realm of human reproduction. This attitude is generally a healthy one, but in some cases it is dangerously misplaced, and "reproductive" cloning is such a case. It is so precisely because "reproductive" cloning would affect more than individuals, and more than the private relationships at the heart of reproductive choices. The way in which the next generation will enter this world has everything to do with the way in which our society will live into the future. A society that produces children through cloning is a society that thinks about children and family and the human condition in a certain way; and we must be given the option, as a society, to decide if that is the way that we wish to think about these most important matters.

Animals Should Be Used in Cloning Research

By Lester M. Crawford

In the following viewpoint, originally given as testimony before the House Subcommittee on Technology on July 22, 1997, Lester M. Crawford, chairman of the National Association for Biomedical Research (NABR) board of directors, implores governmental leaders to support the use of animals in research. He points out that existing animal welfare laws and regulations are sufficient to protect the animals being used in biotechnical research. He emphasizes the great contributions animals have made and could continue to make to the fields of medicine and biotechnology, citing among other things the possibility of animals being cloned as a source of organs for transplantation in humans suffering from heart, liver, or kidney failure. Crawford, also a veterinarian, says cloning research can also benefit animals by leading to improved medical treatments for animals and by uncovering ways to preserve endangered species. Crawford is director of the Center for Food and Nutrition Policy at Georgetown University. He has been a veterinarian for thirty-four years and has served as the executive director of the Association of American Veterinary Medical Colleges.

I am here today as the Chairman-elect of the National Association for Biomedical Research Board of Directors. NABR, as the Association is called, is dedicated exclusively to advocating sound public policy regarding the humane and necessary use of animals in biomedical research, education and testing. NABR represents over 360 distinguished member institutions including

Lester M. Crawford, testimony before the Subcommittee on Technology, House Committee on Science, Washington, DC, July 22, 1997.

the nation's largest universities, the majority of U.S. medical and veterinary schools, academic and professional societies, voluntary health organizations as well as pharmaceutical and biotechnology companies. We appreciate the opportunity to discuss the interests of the animal research community as they pertain to Congress' consideration of legislation concerning human cloning.

Recommendations Made by the National Bioethics Advisory Commission

Like the other scientific organizations from which you have heard, NABR agrees with and supports the conclusions and recommendations made by the National Bioethics Advisory Commission (NBAC) in its June, 1997, report. Let me describe their findings as we understand them. The need to do this underscores the challenge facing this subcommittee and that is to clearly articulate our national policy governing the creation of a human being using somatic cell nuclear transfer. The three central actions that the Commission suggests be taken by government are:

First, attempts to create a human child using the new cloning technology of somatic cell nuclear transfer should not be permitted by anyone in the public or private sector, either in the laboratory or in a clinical setting. This prohibition is recommended not only because the technique currently is medically unsafe to use in humans, but also because there are moral and ethical concerns about this prospect that will likely continue to be deliberated and reviewed well into the future.

Next, the use of somatic cell nuclear transfer in research on cloning animals should continue. This type of research is both acceptable and beneficial to the public. Existing animal welfare laws and regulations, including review by institution-based animal protection committees, are sufficient to address our concerns about animal research.

Likewise, additional limitations should not be placed upon the cloning of human cells and DNA sequences using somatic cell nuclear transfer. These research efforts do not raise the same scientific and ethical issues that surround the possible creation of an entire human being in the laboratory.

Madame Chairwoman, it is understandable that an awesome achievement such as the birth of the lamb called Dolly, the first successful clone from an adult mammal somatic cell, has caused some degree of apprehension. What is not right is for the public's reasonable fears to be exploited by hyperbole. To this end, we in the research community must devote more attention to educating the public. With more reliable information, people will be prepared to separate legitimate science from science fiction. Unfortunately, in addition to some irresponsible, tabloid-type reporting on the subject of cloning, we have seen several radical groups try to recruit Dolly for their own propaganda purposes. One such group staged an anti-animal research demonstration during the NBAC proceedings. The agenda of those few who would stop animal research in any way possible, for any reason whatsoever, must not cloud the important issues before you.

The Essential and Responsible Use of Animals

The Commission supported continuation of the essential and responsible use of animals in biomedical research, and NABR is confident you will do the same in considering legislation regarding human cloning. More than this, NABR believes that the constructive hearing you and the subcommittee are conducting will help alleviate needless apprehension and still encourage the best science. For the public expects research risks to be addressed while research benefits continue.

Great progress in medicine and biotechnology is possible using new genetic techniques without entering the realm of cloning human beings. Genetically engineered mice have already revolutionized our ability to study devastating diseases such as breast cancer and immune system deficiency. Even better animal models for human disease, aiding research into new or improved therapies, are an exciting prospect stemming from the latest cloning methods.

The most immediate benefit is likely to be the faster, more efficient production of therapeutic human proteins in the milk of transgenic [having received genes from another species] farm animal species. These drug products of biotechnology have already aided persons with blood deficiencies and serious infections among other conditions. In the longer-term future, cloned animals might become a safe source of organs for transplantation in patients with heart, kidney or liver failure.

Studying the somatic cell nuclear transfer process itself in animals and human tissue, never approaching the actual cloning of a human, could also provide other long-awaited answers. For example, so-called somatic mutations—mutations that take place in adult human and animal cells that are not inheritable—can cause tumors and other illnesses. Cellular changes of this type are also part of the aging process. Looking at the way cells undergo those sorts of mutations could help us better prevent cancer and avoid the negative effects of growing old, such as Alzheimer's disease. Ultimately, greater understanding of somatic cell differentiation might lead to the ability to regenerate or repair living tissue damaged by a variety of causes including spinal cord injury.

In veterinary and agricultural research, cloning techniques may benefit animals directly either through improved medical treatments or by preserving genetic strengths. Rather than limiting genetic horizons, new technologies may help us to preserve biodiversity and ensure the continuation of rare individual animals or endangered species, too many of which are in need of protection today.

Shared Responsibilities of Science and Government

In summary, NABR believes that science and government have shared responsibilities.

These duties are especially relevant to the national policy question now before you. Together we must reassure the public that:

• science will not pursue research results which society is morally and ethically unwilling to accept;

• research is being facilitated and the rewards of research can be enjoyed because safeguards are in place to protect humans and animals in experimentation; and

• existing laws and regulations are being followed and periodically reviewed to keep pace with new technologies.

Madame Chairwoman, NABR applauds you and the subcommittee for seeking a sound science policy regarding human cloning and trusts that in determining that policy you will promote responsible life-saving research. NABR would be pleased to provide any assistance you may need in the future to be certain that legislative proposals do not impede research requiring animals. Thank you, once again, for the opportunity to express our views and for your interest in these issues. I welcome any questions you may have.

Animals Suffer When Used in Cloning Research

By Michael W. Fox

Researchers in the biomedical field are genetically engineering animals to study human genetic diseases and to produce milk containing pharmaceuticals. In this selection, Michael W. Fox questions the ethics of using animals for these purposes because of the suffering the animals are forced to endure. He says animals are engineered to have genetic and developmental defects in the name of advancing cloning research. Some of the health problems research subjects experience include the development of tumors, chronic kidney and liver dysfunction, and damaged or missing organs. Moreover, he says cloning researchers report high incidences of miscarriages and abnormally large infants. In addition to the suffering humans have inflicted on these animals, Fox believes that researchers have helped humans create an increasingly parasitic relationship with animals. Another concern of Fox's is that cloning animals will lead to genetic uniformity, which will increase species' susceptibility to infectious diseases. Fox is a veterinarian and former vice president of the Humane Society of the United States and the Humane Society International. Fox is also the author of over forty books on animal care and behavior.

In March 1997, the news media around the world splashed the story of a British scientist, Ian Wilmut, who had succeeded in making the first clone of a mammal from an adult cell. Wilmut used a cell taken from the udder tissue of a sheep fused into an empty ovum or egg, which was then implanted into the womb of a surrogate ewe. This replica, called Dolly, along with her creator, became an instant international celebrity. Some likened Wilmut's achievement to those of Galileo, Einstein, and Copernicus. But cloning is no worldview-changing discovery. Other nonmammalian creatures such as frogs have been cloned by biologists in the past, and for the vast majority of life forms on earth, it is the asexual way of achieving species survival and multiplication. This new technology, however, was so newsworthy because it is disturbing to the public, bringing us to the threshold of human cloning—which has already been done with human embryos by scientists in South Korea (and probably in other laboratories in other countries), but embryos were only developed in vitro and not allowed to grow more than a few cells. Critics linked this new technique with Mary Shelley's *Frankenstein* story, reporters asked if Dolly has a soul, and three out of four Americans in a *Time*/CNN survey said they believed such research is "against the will of God." Significantly, perhaps, when asked by a reporter at a 1997 Senate subcommittee hearing on cloning if he held any particular religious belief, Dr. Wilmut said that he categorized himself as an agnostic with no particular religious belief or theology.

What is the point in cloning animals, and where might it all lead? One immediate fear is that humans will be cloned once the technique is perfected. Only 29 of some 277 sheep embryos cloned by Wilmut developed normally. Many died before birth, had defective kidneys, or were abnormally large—not the carbon-copy replicas the biotechnologist had anticipated. Concern about the welfare of such animals, as with transgenic animals [animals who have genes from another species], is neither effectively regulated nor addressed in most research reports.

The U.K.'s Roslin Institute, where Dolly was created, quickly moved to patent its proven cloning technique. Rural Advancement Foundation International comments as follows on the Roslin Institute's world patent on the cloning of all animal species, including humans:

> The UK's Roslin Institute is so sure it has an economic winner it is claiming its cloning patents in even the weakest of economies—North Korea and Liberia, for instance. The patents are licensed to PPL [Pharmaceutical Proteins, Ltd.] Therapeutics, a company which has agreements with major drug multinationals like Novo Nordisk, Boehringer Ingleheim, and American Home Products. More licenses may be granted. Unlike many bioengineering

patents, which are specified for "nonhumans," Roslin says its cloning patents cover all animals, including humans.

Soon after the news splash about Dolly, other sheep, cows, pigs, and monkeys were reported to be pregnant with clones created by methods similar to those used to make Dolly the sheep. Several researchers reported high embryo mortality (miscarriages) and abnormally large size. Veterinarian Mark Westhusin at Texas A&M University stated, "We've had some [calves] that were just monstrous, up to 180 pounds," compared with the normal 80 pounds. Oregon Primate Research Center reported creating two rhesus monkey clones and had several monkeys pregnant with clones. Scientists in many fields, including AIDS research, have already put in requests for such monkeys.

Cloned Animals Suffer

The specter of cloned animal suffering is, however, very real. Dolly-like clones at the Roslin Institute that died soon after birth were larger than normal, putting their mothers at risk, and had congenital abnormalities in their kidneys and cardiovascular systems—problems not reported in the scientists' first paper in *Nature* magazine. Miscarriages may be due to improper development of the placenta. In recent studies of Dolly, Dr. Wilmut has found subtle changes in chromosome structure usually found in cells of older animals—evidence of a possible "molecular memory," as the cell that was used to create Dolly came from a six-year-old sheep.

Likewise, it was some time after the creation of giant "supermice" by R.D. Palmiter and coworkers that many health problems were reported, notably chronic kidney and liver dysfunction; tumor development; damage to female reproductive organs; structural changes in the heart, spleen, and salivary glands; plus shorter life spans and high infant and juvenile mortality.

In a press release from Tass News Agency, Moscow, over one hundred new types of animals have been cloned in Russia, opening "boundless opportunities foremost in intensive industrial cattle-breeding," as well as pig organs for humans and new varieties of sheep that have biopharmaceuticals in their milk to treat gastrointestinal diseases and more quickly ferment sheep milk into cheese. China has reported cloning of pigs, rabbits, and cattle, and researchers in Australia and Denmark are developing cow clones.

In September 1997, PPL Therapeutics of Edinburgh, a corporate offshoot of the creators of Dolly, produced a sheep called Polly. She was a clone like Dolly but was also the first transgenic animal to be cloned; she bore a human gene that would cause her to produce alpha-1-antitrypsin, a human blood protein used to treat cystic fibrosis, in her milk. Of two other live transgenic clones produced at the same time as Polly, from a total of sixty-two embryos implanted into surrogate mothers, one died soon after birth. Also, some births had to be induced or performed by Caesarian section because, for unknown reasons, natural labor failed to occur, according to a *Washington Post* report.

PPL's American Division in Blacksburg, Virginia, soon after announced its success in creating three transgenic rabbits, whose milk contains calcitonin (from an inserted salmon gene), a bone-building substance that can be used to treat osteoporosis. The company plans to try this procedure in farm animals. Around this same time, ABS Global, Inc., of DeForest, Wisconsin, unveiled a healthy six-month-old calf named Gene developed with its own patented cloning technology.

The *Veterinary Record* of October 18, 1997, reported that scientists at the Roslin Institute are now particularly interested in creating in sheep the same range of mutations that give rise to cystic fibrosis in humans. Asked whether he felt it was morally correct to introduce genetic defects into an animal, Dr. Wilmut (Dolly's creator) said that, provided the animals receive the same degree of care as a person with the disease and there was otherwise no realistic prospect of bringing forward treatment, he would be "quite comfortable" to be involved in such work.

On a BBC TV *Horizon* documentary entitled "Hello, Dolly," the founders of Granada Genetics in Texas, who were the first to market cloned cattle, admitted that their venture had failed because many bovine clones were abnormally large and had to be delivered by Caesarian section, had enlarged hearts, and developed diabetes.

Scientists at the University of Hawaii announced in 1998 the birth of a third generation of mice cloned from adult cells, demonstrating the most successful technique to date. But many gene engineers insist the technology still has far to go before it is commercially viable, with only a 1 to 2 percent success rate of eggs injected with an adult animal's cell. An article in *Nature Biotechnology* included the following:

"While we know how to do nuclear transfer with adult somatic cells, we don't know how to do it efficiently," said Ian Wilmut at the recent Second Annual Congress on Mammalian Cloning in Washington, DC. . . . "There are losses at every stage of development of the embryos, as well as serious congenital malformations in many animals."

This was echoed by a number of speakers at the meeting, who noted that "a great number" of animals born from nuclear transfer have been born missing organs such as kidneys and hearts, and that many have also been stillborn or born greatly oversized. "Through our current methods of nuclear transfer, we don't really know what kind of embryos we are creating—normal or not," said Tanja Dominko, staff scientist in Gerald Schatten's laboratory, at the Oregon Regional Primate Research Center (Beaverton), and who used to work on bovine cloning at the University of Wisconsin.

Dominko and Dr. Neal L. First of the University of Wisconsin have emptied out cow ova and reportedly succeeded in getting cells from the ears of mature pigs, sheep, and monkeys to become embryonic inside these ova. None implanted, however, when put into the uterus of a surrogate animal mother.

Cloning Human Organs

Dr. First had hoped this technique could provide a way to clone endangered species, but now other biotechnologists are developing this technique of using emptied-out cows' eggs to contain fused human embryo cells, called stem cells. The aim is to clone various human organs, as well as brain and heart muscle cells, for transplant into humans. Patients' own cells could also be cloned to reduce chances of immune rejection, according to Jose Cibelli of Advanced Cell Technology in Worcester, Massachusetts. Viable human embryos would not be produced in this process, so one ethical hurdle would be avoided. But the use of cows as surrogates in the production of spare parts for humans raises the specter of an increasingly parasitic human relationship with and dependence on animals that is aesthetically disturbing, if not biologically regressive.

A new way possibly to create spare organ parts has been described by Dr. Jonathan Slack of Bristol University. Slack has controlled frog egg genes to create headless tadpoles. He says his research could lead to the production of cultured human organs such as a heart or pancreas in reprogrammed, cloned human eggs. Conceivably headless humans, like his headless tadpoles, could be cloned, but with perfection of this technology, specific body parts could be created from cloned human cells.

A new technique using calf embryonic stem cells to create clones was announced in June 1998 by researchers at the University of Massachusetts, who produced twelve calf clones. Their company, Advanced Cell Technology, Inc., has contracted with Genzyme Transgenics to develop genetically engineered cows that produce in their milk a human protein that is used to treat blood loss, using the cloning technique to rapidly build up a uniform herd of transgenic animals. France has also joined the cloning craze, showing a video of "Margueritte," a calf cloned from a muscle cell of a calf fetus, at the Paris Agricultural Show in March 1998.

In the fall of 1998 the biggest set of clones so far from one animal at one time were produced—ten calves from cultured cells of one cow were born in New Zealand at the government's Agricultural Research Center in Ruakura. Around the same time, Snow Brand Milk Products (SBMP Hokkaido), one of Japan's largest dairy companies, succeeded in impregnating cows with embryos cloned from mammary gland cells extracted from cow's milk. If successfully born, the calves would be the first animals cloned from somatic cells collected from milk.

One real problem with creating animal clones is their genetic uniformity, which is likely to decrease their survivability by increasing their vulnerability to infectious diseases. Another problem is the suffering of those animals afflicted by genetic and developmental defects caused accidentally because the technology is not risk free or caused deliberately for biomedical research into human genetic and developmental diseases.

If cloning biotechnology is ever perfected, what will it mean for humans and other animals? The technique developed by Wilmut has been patented, so venture capitalists have high hopes that cloning will be a boost to the organ transplant industry and to pharmaceutical "pharming" of health care products. Pigs, sheep, cattle, and goats have already been genetically engineered, respectively, to serve as organ donors for people, to produce more humanized milk, and to produce milk containing valuable biopharmaceuticals. The numbers of these animals might now be rapidly increased using cloning biotechnology.

CHAPTER 4

Further Cloning Developments

The Future of Cloning Research in the United States

By Robert A. Weinberg

Molecular biologist Robert A. Weinberg explains in this selection that biologists believe that reproductive cloning technology is riddled with possibly unsolvable problems. Weinberg believes these problems create a dead end for reproductive cloning (used to produce genetically identical offspring). He suggests that there is little chance that humans will be cloned except in the minds of scientists who have little credibility and are participating in what Weinberg calls the "cloning circus." In the cloning circus would-be cloners mainly interested in making money send their experiment results directly to the mainstream press or lesser-known biotech publications where the usual quality control exercised by a scientific peer-review process is bypassed. The peer-review process ensures the credibility of new discoveries, and without it, Weinberg claims, the scientific authenticity of experiments is lost. As a result of the cloning circus, all cloning research in the United States will be harmed, including therapeutic cloning research (used to find cures for human diseases), which, unlike reproductive cloning research, shows great promise.

Robert A. Weinberg is a professor of biology at the Massachusetts Institute of Technology. He is the author of several books on cancer research including *Racing to the Beginning of the Road* and *One Renegade Cell*. Weinberg is the winner of the 1997 National Medal of Science, and is a member of the Whitehead Institute for Biomedical Research.

Robert A. Weinberg, "Of Clones and Clowns," *The Atlantic Monthly*, vol. 289, June 2002, pp. 55–59.

Biologists have been rather silent on the subject of human cloning. Some others would accuse us, as they have with predictable regularity in the recent past, of insensitivity to the societal consequences of our research. If not insensitivity, then moral obtuseness, and if not that, then arrogance—an accusation that can never be disproved.

The truth is that most of us have remained quiet for quite another reason. Most of us regard reproductive cloning—a procedure used to produce an entire new organism from one cell of an adult—as a technology riddled with problems. Why should we waste time agonizing about something that is far removed from practical utility, and may forever remain so?

The nature and magnitude of the problems were suggested by the Scottish scientist Ian Wilmut's initial report, [in 1997], on the cloning of Dolly the sheep. Dolly represented one success among 277 attempts to produce a viable, healthy newborn. Most attempts at cloning other animal species—to date cloning has succeeded with sheep, mice, cattle, goats, cats, and pigs—have not fared much better.

Even the successes come with problems. The placentas of cloned fetuses are routinely two or three times larger than normal. The offspring are usually larger than normal as well. Several months after birth one group of cloned mice weighed 72 percent more than mice created through normal reproduction. In many species cloned fetuses must be delivered by cesarean section because of their size. This abnormality, the reasons for which no one understands, is so common that it now has its own name—Large Offspring Syndrome. Dolly (who was of normal size at birth) was briefly overweight in her young years and suffers from early-onset arthritis of unknown cause. Two recent reports indicate that cloned mice suffer early-onset obesity and early death.

Arguably the most successful reproductive-cloning experiment was reported [in 2001] by Advanced Cell Technology (ACT), a small biotech company in Worcester, Massachusetts. Working with cows, ACT produced 496 embryos by injecting nuclei from adult cells into eggs that had been stripped of their own nuclei. Implanting the embryos into the uteruses of cows led to 110 established pregnancies, thirty of which went to term. Five of the newborns died shortly after birth, and a sixth died several months later. The twenty-four surviving calves developed into cows that were healthy by all criteria examined. But most, if not all, had enlarged placentas, and as newborns some of them suffered from the respiratory distress typical of Large Offspring Syndrome.

The success rate of the procedure, roughly five percent, was much higher than the rates achieved with other mammalian species, and the experiment was considered a great success. Some of the cows have grown up, been artificially inseminated, and given birth to normal offspring. Whether they are affected by any of the symptoms associated with Large Offspring Syndrome later in life is not apparent from the published data. No matter: for $20,000 ACT will clone your favorite cow.

Imagine the application of this technology to human beings. Suppose that 100 adult nuclei are obtained, each of which is injected into a human egg whose own nucleus has been removed. Imagine then that only five of the 100 embryos thus created result in well-formed, viable newborns; the other ninety-five spontaneously abort at various stages of development or, if cloning experiments with mammals other than cows are any guide, yield grossly malformed babies. The five viable babies have a reasonable likelihood of suffering from Large Offspring Syndrome. How they will develop, physically and cognitively, is anyone's guess. It seems unlikely that even the richest and most egomaniacal among us, intent on recreating themselves exactly, will swarm to this technology.

The Importance of the Peer-Review Process

Biological systems are extraordinarily complex, and there are myriad ways in which experiments can go awry or their results can be misinterpreted. Still, perhaps 95 percent of what biologists read in . . . research journals will be considered valid (if perhaps not very interesting) a century from now. Much of scientists' trust in the existing knowledge base derives from the system constructed over the past century to validate new research findings and the conclusions derived from them. Research journals impose quality controls to ensure that scientific observations and conclusions are solid and credible. They sift the scientific wheat from the chaff.

The system works like this: A biologist sends a manuscript describing his experiment to a journal. The editor of the journal recruits several experts, who remain anonymous to the researcher, to vet the manuscript. A month or two later the researcher receives a thumbs-up, a thumbs-down, or a request for revisions and more data. The system works reasonably well, which is why many of us invest large amounts of time in serving as the anonymous reviewers of one another's work. Without such rigorously imposed quality control, our subfields of research would rapidly descend into chaos, because no publicly announced result would carry the imprimatur of having been critiqued by experts.

We participate in the peer-review process not only to create a sound edifice of ideas and results for ourselves; we do it for the outside world as well—for all those who are unfamiliar with the arcane details of our field. Without the trial-by-fire of peer review, how can journalists and the public possibly know which discoveries are credible, which are nothing more than acts of self-promotion by ambitious researchers, and which smack of the delusional?

The hype about cloning has made a shambles of this system, creating something of a circus. Many of us have the queasy feeling that our carefully constructed world of science is under siege. The clowns—those who think that making money, lots of it, is more important than doing serious science—have invaded our sanctuary.

The Cloning Circus

The cloning circus opened soon after Wilmut, a careful and well-respected scientist, reported his success with Dolly. First in the ring was Richard Seed, an elderly Chicago physicist, who in late 1997 announced his intention of cloning a human being within two years. Soon members of an international religious cult, the Raëlians (followers of Claude Vorilhon, a French-born mystic who says that he was given the name Raël by four-foot-high extraterrestrials, and who preaches that human beings were originally created by these aliens), revealed an even more grandiose vision of human cloning. To the Raëlians, biomedical science is a sacrament to be used for achieving immortality: their ultimate goal is to use cloning to create empty shells into which people's souls can be transferred. As a sideline, the Raëlian-affiliated company Clonaid hopes to offer its services to couples who would like to create a child through reproductive cloning, for $200,000 per child.

Neither Seed nor the Raëlians made any pretense of subjecting their plans to review by knowledgeable scientists; they went straight to the popular press. Still this wasn't so bad. Few science journalists took them seriously (although they did oblige them with extensive coverage). Biologists were also unmoved. Wasn't it obvious that Seed and the Raëlians were unqualified to undertake even the beginnings of the series of technical steps required for reproductive cloning? Why dignify them with a response?

The next wave of would-be cloners likewise went straight to the mainstream press—but they were not so easily dismissed. In March of 2001, at a widely covered press conference in Rome, an Italian and a U.S. physician announced plans to undertake human reproductive cloning outside the United States. The Italian member of the team was Severino Antinori, a gynecologist notorious for having used donor eggs and *in vitro* fertilization to make a sixty-two-year-old woman pregnant in 1994. Now he was moving on. Why, he asked, did the desires of infertile couples (he claimed to have 600 on a waiting list) not outweigh the concerns about human cloning? He repeatedly shouted down reporters and visiting researchers who had the temerity to voice questions about the biological and ethical problems associated with reproductive cloning.

The American member of the team was Panayiotis Zavos, a reproductive physiologist and an *in vitro* fertilization expert at the Andrology Institute of America, in Lexington, Kentucky. "The genie is out of the bottle," he told reporters. "Dolly is here, and we are next." Antinori and Zavos announced their intention of starting a human cloning project in an undisclosed Mediterranean country. Next up was Avi Ben-Abraham, an Israeli-American biotechnologist with thwarted political ambitions (he ran unsuccessfully for the Knesset) and no reputable scientific credentials, who attempted to attach himself to the project. Ben-Abraham hinted that the work would be done either in Israel or in an Arab country, because "the climate is more [receptive to human cloning research] within Judaism and Islam." He told the German magazine *Der Spiegel*, "We were all created by the Almighty, but now we will become the creators."

Both Antinori and Zavos glossed over the large gap between expertise with established infertility procedures and the technical skills required for reproductive cloning. Confronted with the prospect of high rates of aborted or malformed cloned embryos, they claimed to be able to weed out any defective embryos at an early stage of gestation. "We have a great deal of knowledge," Zavos announced to the press. "We can grade embryos. We can do genetic screening. We can do [genetic] quality control." This was possible, he said, because of highly sensitive diagnostic tests that can determine whether or not development is proceeding normally.

The fact is that no such tests exist; they have eluded even the most expert biologists in the field, and there is no hope that they will be devised anytime soon—if ever. No one knows how to determine with precision whether the repertoire of genes expressed at various stages of embryonic development is being "read" properly in each cell type within an embryo. Without such information, no one can know whether the developmental program is proceeding normally in the womb. (The prenatal tests currently done for Down syndrome and several other genetic disorders can detect only a few of the thousands of things that can go wrong during embryonic development.)

Rudolf Jaenisch, a colleague of mine with extensive experience in mouse reproductive cloning, was sufficiently exercised to say to a reporter at the *Chicago Tribune*, "[Zavos and Antinori] will produce clones, and most of these will die in utero. . . . Those will be the lucky ones. Many of those that survive will have [obvious or more subtle] abnormalities." The rest of us biologists remained quiet. To us, Antinori, Zavos, and Ben-Abraham were so clearly inept that comment seemed gratuitous. In this instance we have, as on other occasions, misjudged the situation: many people seem to take these three and their plans very seriously indeed. And, in fact, [in April 2002], Antinori claimed, somewhat dubiously, that a woman under his care was eight weeks pregnant with a cloned embryo.

Therapeutic Cloning and the Biotech Industry

In the meantime, the biotechnology industry, led by ACT, has been moving ahead aggressively with human cloning, but of a different sort. The young companies in this sector have sensed, probably correctly, the enormous potential of therapeutic (rather than reproductive) cloning as a strategy for treating a host of common human degenerative diseases.

The initial steps of therapeutic cloning are identical to those of reproductive cloning: cells are prepared from an adult tissue, their nuclei are extracted, and each nucleus is introduced into a human egg, which is allowed to develop. However, in therapeutic cloning embryonic development is halted at a very early stage—when the embryo is a blastocyst, consisting of perhaps 150 cells—and the inner cells are harvested and cultured. These cells, often termed embryonic stem cells, are still very primitive and thus have retained the ability to develop into any type of cell in the body (except those of the placenta).

Mouse and human embryonic stem cells can be propagated in a petri dish and induced to form precursors of blood-forming cells, or of the insulin-producing cells of the pancreas, or of cardiac muscle or nerve tissue. These precursor cells (tissue-specific stem cells) might then be introduced into a tissue that has grown weak from the loss of too many of its differentiated worker cells. When the ranks of the workers are replenished, the course of disease may be dramatically reversed. At least, that is the current theory. . . . One version of the technique has been successfully applied to mice.

Therapeutic cloning has the potential to revolutionize the treatment of a number of currently untreatable degenerative diseases, but it is only a potential. Considerable research will be required to determine the technology's possibilities and limitations for treating human patients.

Some worry that therapeutic-cloning research will never get off the ground in this country. Its proponents—and there are many among the community of biomedical researchers—fear that the two very different kinds of cloning, therapeutic and reproductive, have merged in the public's mind. Three leaders of the community wrote a broadside . . . in *Science*, titled "Please Don't Call It Cloning!" Call therapeutic cloning anything else—call it "nuclear transplantation," or "stem cell research." The scientific community has finally awakened to the damage that the clowns have done.

The Government Stops Human Embryonic Stem Cell Research

This is where the newest acts of the circus begin. President George [W.] Bush and many pro-life activists are in one ring. A number of disease-specific advocacy groups that view therapeutic cloning as the only real prospect for treating long-resistant maladies are in another. In a third ring are several biotech companies that are flogging their wares, often in ways that make many biologists shudder.

Yielding to pressure from religious conservatives, Bush announced [in] August [2001] that no new human embryonic stem cells could be produced from early human embryos that had been created during the course of research sponsored by the federal government: any research on the potential applications of human embryonic stem cells, he said, would have to be conducted with the existing repertoire of sixty-odd lines. The number of available, usable cell lines actually appears to be closer to a dozen or two. And like all biological reagents, these cells tend to deteriorate with time in culture; new ones will have to be derived if research is to continue. What if experiments with the existing embryonic-stem-cell lines show enormous promise? Such an outcome would produce an almost irresistible pressure to move ahead with the derivation of new embryonic stem cells and to rapidly expand this avenue of research.

How will we learn whether human embryonic stem cells are truly useful for new types of therapy? This question brings us directly to another pitfall: much of the research on human embryonic stem cells is already being conducted by biotech companies, rather than in universities. Bush's edict will only exacerbate this situation. (In the 1970s a federal decision effectively banning government funding of *in vitro* fertilization had a similar effect, driving such research into private clinics.)

Biotech Companies Bypass Peer-Review Process

Evaluating the science coming from the labs of the biotech industry is often tricky. Those who run these companies are generally motivated more by a need to please stock analysts and venture capitalists than to convince scientific peers. For many biotech companies the peer-review process conducted by scientific journals is simply an inconvenient, time-wasting impediment. So some of the companies routinely bypass peer review and go straight to the mainstream press. Science journalists, always eager for scoops, don't necessarily feel compelled to consult experts about the credibility of industry press releases. And when experts are consulted about the contents of a press release, they are often hampered by spotty descriptions of the claimed breakthrough and thus limited to mumbling platitudes.

ACT, the company that conducted the successful cow-cloning experiment and has now taken the lead in researching human therapeutic cloning, has danced back and forth between publishing in respectable peer-reviewed journals and going directly to the popular press—and recently tried to find a middle ground. [In the fall of 2001], with vast ambitions, ACT reported that it had conducted the first successful human-cloning experiment. In truth, however, embryonic development went only as far as six cells—far short of the 150-cell blastocyst that represents the first essential step of therapeutic cloning. Wishing to cloak its work in scientific respectability, ACT reported these results in a fledgling electronic research journal named *e-biomed: The Journal of Regenerative Medicine.* Perhaps ACT felt especially welcome in a journal that, according to its editor in chief, William A. Haseltine, a widely known biotech tycoon, "is prepared to publish work of a more preliminary nature." It may also have been encouraged by Haseltine's stance toward cloning, as revealed in his remarks when the journal was founded. "As we understand the body's repair process at the genetic level, we will be able to advance the goal of maintaining our bodies in normal function, perhaps perpetually," he said.

Electronic publishing is still in its infancy, and the publication of ACT's research report will do little to enhance its reputation. By the usual standards of scientific achievement, the experiments ACT published would be considered abject failures. Knowledgeable readers of the report were unable to tell whether the clump of six cells represented the beginning of a human embryo or simply an unformed aggregate of dying cells.

One prominent member of the *e-biomed* editorial board, a specialist in the type of embryology used in cloning, asked Haseltine how the ACT manuscript had been vetted before its publication. Haseltine assured his board member that the paper had been seen by two competent reviewers, but he refused to provide more details. The board member promptly resigned. Two others on the editorial board, also respected embryologists, soon followed suit. (Among the scientists left on the board are two representatives of ACT—indeed, both were authors of the paper.) Mary Ann Liebert, the publisher of the journal, interpreted this exodus as a sign that "clearly some noses were out of joint." The entire publication process subverted the potentially adversarial but necessary dynamic between journal-based peer review and the research scientist.

The Promise of Adult Stem Cells

No one yet knows precisely how to make therapeutic cloning work, or which of its many claimed potential applications will pan out and which will not. And an obstacle other than experimental problems confronts those pushing therapeutic cloning. In the wake of the cloning revolution a second revolution has taken place—

quieter but no less consequential. It, too, concerns tissue-specific stem cells—but ones found in the tissues of adults. These adult stem cells may one day prove to be at least as useful as those generated by therapeutic cloning.

Many of our tissues are continually jettisoning old, worn-out cells and replacing them with freshly minted ones. The process depends on a cadre of stem cells residing in each type of tissue and specific to that type of tissue. When an adult stem cell divides, one of its two daughters becomes a precursor of a specialized worker cell, able to help replenish the pool of worker cells that may have been damaged through injury or long-term use. The other remains a stem cell like its mother, thus ensuring that the population of stem cells in the tissue is never depleted.

Until [the year 2000] the dogma among biologists was that stem cells in the bone marrow spawned only blood, those in the liver spawned only hepatocytes, and those in the brain spawned only neurons—in other words, each of our tissues had only its own cadre of stem cells for upkeep. Once again we appear to have been wrong. There is mounting evidence that the body contains some rather unspecialized stem cells, which wander around ready to help many sorts of tissue regenerate their worker cells.

Whether these newly discovered, multi-talented adult stem cells present a viable alternative to therapeutic cloning remains to be proved. Many of the claims about their capabilities have yet to be subjected to rigorous testing. Perhaps not surprisingly, some of these claims have also reached the public without careful vetting by peers. Senator Sam Brownback, of Kansas, an ardent foe of all kinds of cloning, has based much of his case in favor of adult stem cells (and against therapeutic cloning) on these essentially unsubstantiated scientific claims. Adult stem cells provide a convenient escape hatch for Brownback. Their use placates religious conservatives, who are against all cloning, while throwing a bone to groups lobbying for new stem-cell-based therapies to treat degenerative diseases.

Brownback would have biologists shut down therapeutic-cloning research and focus their energies exclusively on adult stem-cell research. But no one can know at present which of those two strategies is more likely to work. It will take a decade or more to find out. Many biologists are understandably reluctant to set aside therapeutic-cloning research in the meantime; they argue that the two technologies should be explored simultaneously.

Precisely this issue was debated recently by advisory committees in the United States and Germany. The U.S. committee was convened by Bruce Alberts, the president of the National Academy of Sciences [NAS] and a highly accomplished cell biologist and scientific educator. Quite naturally, it included a number of experts who are actively involved in exploring the advantages and disadvantages of stem-cell therapies. The committee, which announced its findings in January [2002], concluded that therapeutic cloning should be explored in parallel with alternative strategies.

For their trouble, the scientists were accused of financial self-interest by Steven Milloy of Fox News, who said, "Enron and Arthur Andersen [which were accused of financial malfeasance] have nothing over the National Academy of Sciences when it comes to deceiving the public. . . . Enter Bruce Alberts, the Wizard of Oz–like president of the NAS. . . . On his own initiative, Alberts put together a special panel, stacked with embryonic-stem-cell research proponents and researchers already on the taxpayer dole. . . . Breast-feeding off taxpayers is as natural to the NAS panel members as breathing."

The German committee, which reached a similar conclusion, was assembled by Ernst-Ludwig Winnacker, the head of his country's national science foundation. Winnacker and his colleagues were labeled "cannibals" by the Cardinal of Cologne. Remarks like the ones from Steven Milloy and the cardinal seem calculated to make public service at the interface between science and society as unappealing as possible.

President Bush, apparently anticipating the NAS panel's conclusion, has appointed an advisory committee all but guaranteed to produce a report much more to his liking. Its chairman, Leon Kass, has gone on record as being against all forms of cloning. (Earlier in his career Kass helped to launch an attack on *in vitro* fertilization.)

Meanwhile, a coalition of a hundred people and organizations recently sent a letter to Congress expressing their opposition to therapeutic cloning—among them Friends of the Earth, Greenpeace, the Sierra Club, the head of the National Latina Health Organization, and the perennial naysayer Jeremy Rifkin. "The problem with therapeutic cloning," Rifkin has said, "is that it introduces commercial eugenics from the get-go." Powerful words indeed. Few of those galvanized by Rifkin would know that therapeutic cloning has nothing whatsoever to do with eugenics.

Politics Will Settle the Debate

Usually progress in biology is held back by experimental difficulties, inadequate instruments, poorly planned research protocols, inadequate funding, or plain sloppiness. But in this case the future of research may have little connection with these factors or with the scientific pros and cons being debated earnestly by members of the research community. The other, more public debates will surely be the decisive ones.

The clashes about human therapeutic cloning that have taken place in the media and in Congress are invariably built around weighty moral and ethical principles. But none of us needs a degree in bioethics to find the bottom line in the arguments. They all ultimately converge on a single question: When does human life begin? Some say it is when sperm and egg meet, others when the embryo implants in the womb, others when the fetus quickens, and yet others when the fetus can survive outside the womb. This is a question that we scientists are neither more nor less equipped to decide than the average man or woman in the street, than a senator from Kansas or a cardinal in Cologne. (Because Dolly and the other cloned animals show that a complete embryo can be produced from a single adult cell, some biologists have proposed, tongue in cheek, that a human life exists in each one of our cells.) Take your pick of the possible answers and erect your own moral scaffolding above your choice.

In the end, politics will settle the debate in this country about whether human therapeutic cloning is allowed to proceed. If the decision is yes, then we will continue to lead the world in a crucial, cutting-edge area of biomedical research. If it is no, U.S. biologists will need to undertake hegiras to laboratories in Australia, Japan, Israel, and certain countries in Europe—an outcome that would leave American science greatly diminished.

A Claim That Humans Have Been Cloned Threatens to Derail Cloning Research

By Nancy Gibbs

According to Nancy Gibbs in the following selection, at the beginning of 2003, chemist Brigitte Boisselier, president of the biotech
company Clonaid, claimed to have cloned a human baby named Eve.
Boisselier belongs to the religious cult called the Raelians, who believe humans descended from clones placed on the earth by space
aliens twenty-five thousand years ago. Scientists and bioethicists re-
acted with disgust at the news, reports Gibbs. Although it is generally accepted that human reproductive cloning (cloning to produce
a child) should be banned, activists believe that Boisselier's announcement is a setback for therapeutic cloning, which some believe
will help cure diseases. Supporters of a total ban on all cloning are
capitalizing on the fear created by the Clonaid news to boost their
cause. In addition, according to Gibbs, desperate prospective parents
are expected to be asking for Clonaid's services, although Boisselier
has provided no proof of baby Eve's existence. Nancy Gibbs is a senior editor for *Time* magazine.

Nancy Gibbs, "Abducting the Cloning Debate," *Time*, vol. 161, January 13, 2003, pp. 46–49.
Copyright © 2003 by Time, Inc. Reprinted by permission.

We think of science as a clean and logical place where, with the right skills and instruments, you can see the world in a grain of sand. So what happens when you cross science with a circus full of clowns and tricks and gaudy lights, where everything is for sale and nothing is for real?

The science circus comes to town when a group like the Raelians claims to be cloning children, announcing one arrival just in time to fill the holiday news vacuum. The news came as a shock but not much of a surprise. It was only a matter of time before one of the teams racing to produce the first human clone either succeeded or just decided to claim it had. Chemist Brigitte Boisselier, president of the biotech company Clonaid, is a member of the Order of Angels of the Raelian religious cult, whose prophet Rael says 4-ft.-tall green space aliens visited him 30 years ago in a French volcano and revealed that all of us are descended from the clones they planted here 25,000 years ago. With her announcement of a miracle baby named Eve and the group's subsequent claim of a second cloned birth, the most important debate in morals and medicine is delivered into such hands to mangle.

Activists who argue passionately over the ethics of cloning, and research on embryos in general, found themselves united in their disgust. When he saw the Raelians on TV, says Arthur Caplan, director of the Center for Bioethics at the University of Pennsylvania, he thought, "Preposterous announcement by kooks." But he also felt despair, as did many scientists who believe that the only way the most morally intricate research can proceed is by keeping it away from charlatans. "I knew they would have a very damaging impact on the cloning debate. First, they would just plain scare people," Caplan notes. "The Raelians are not the picture you want in people's minds when they write their Congressman about cloning."

And write they will, as Congress returns to wrestle with where to erect the guardrails around new reproductive technologies. There is a near consensus for outlawing what the Raelians claim to be doing—cloning one person's cells in order to grow a genetic replica—on the grounds that the risks are too great and the moral costs too high. But so far, no national ban has been passed because a fierce debate still surrounds other forms of research that borrow some of the same techniques. Supporters of "therapeutic cloning," in which embryos are cloned to harvest their stem cells but never grown into a baby, argue that these primitive cells, which can turn

into any kind of cell in the body, may hold the secret to cures for
Parkinson's, Alzheimer's and other diseases. "Of course, all society—from scientists to politicians—is against human reproductive cloning," asserts Dr. Robert Lanza, medical director of Advanced Cell Technology, a biotech firm in Worcester, Mass., that
has led the way in cloning human embryos for stem-cell research.
"No one wants to see 100 copies of Madonna or Michael Jordan.
But it would be tragic if this outrage spills over into legitimate
medical research that could cure millions of patients."

It is harder for the biotech companies to argue for compromise
in a world where the worst-case scenario is getting all the attention. The Raelians are to the labs of America what Enron [a company accused of financial malfeasance] was to the boardrooms, a
rebuke to the premise that science can be self-policing. "If you allow embryo cloning in research labs because of its supposed great
potential," argues Representative Dave Weldon, Republican from
Florida who did research in molecular genetics in graduate school,
"you're going to have all these labs with all these embryos, and it
will be that much easier for people like the Raelians to try to do
reproductive cloning.". . . Congress passed a bill banning all
cloning, but it died in the Senate, where lawmakers still hoped to
write rules that will allow some embryonic research to proceed.
Thanks to the Baby Eve announcement, supporters of a total ban
feel that their chances are now much improved: "I think that gave
it more of a sense of a clear and present danger," says Kansas Senator Sam Brownback, who plans to reintroduce his bill quickly.
His legislation would make it illegal even to import products derived from cloning done overseas—which some say raises the possibility that if scientists in Britain find a cure for Alzheimer's,
American patients will be barred from getting it.

Indeed, some scientists argue that a total cloning ban would im-
pel top U.S. scientists to move overseas, where there is more public support. Britain banned reproductive cloning but is allowing
therapeutic research to move forward. "Blanket bans on technology are almost always a mistake," argues Tim Caulfield, research
director of the Health Law Institute at the University of Alberta.
"You don't ban fertilizer because you can use it to make bombs.
Don't ban cloning because it may be abused. What we should do
is regulate the activities that may be abused, like human reproductive cloning."

No one doubts there's a demand for human cloning. On its website, Clonaid estimates it will charge $200,000 for its reproductive service, but Boisselier insisted to *Time* that so far she has not charged the first guinea pigs. Clonaid also sells human eggs for about $5,000 each and offers "banks" in which to store cells in case a family wants to clone a loved one in the future. Boisselier also has a pet-cloning service called Clonapet, which she says has also received great interest. "The media only want to talk about possible birth defects, that the baby will be a monster, but the e-mails I get from people tell us we're brave, that we should go ahead," says Boisselier.

Still, there was ample reason to challenge her claims of success, even before she began backing off her promise of providing proof. No one has yet succeeded in cloning a primate despite thousands of tries; efforts at a dog have so far failed as well. Even among other mammals, more than 90% of the embryos never implant or die before or soon after birth. Among those most dismissive of her entire operation are the other researchers rushing to beat it, such as Italian fertility specialist Severino Antinori. He missed his own deadline, having announced [in the spring of 2002] that he had a clone due in November. But he is quite certain Boisselier, who has yet to produce Baby Eve, hasn't succeeded either. "It's a great bluff," he barked into his cell phone. "I'm amazed the media believe this."

The damage is done whether Clonaid's claims are a hoax or not. The Raelians can be assured that all the free advertising has worked, and inquiries from prospective parents will rise with each new headline. This desperation leads some lawmakers, ethicists and scientists themselves to argue that it is time to take a broader look at the rules that govern reproductive science. According to a new survey by Johns Hopkins University, two-thirds of Americans approve of using genetic screening to help parents have a baby free of a serious genetic disorder. But more than 70% are against using such techniques to design children to be smarter or more attractive, and 76% are against working on ways to clone humans.

So what should be permissible and what should not? Does the promise of a new technology outweigh the risks that it could be misused? The challenges are too important to address in a climate of fear or ignorance or to be distorted by the greed or vainglory of renegade scientists with an alien agenda.

After Dolly: Problems with Cloning Animals

By Elizabeth Pennisi and Gretchen Vogel

Scientists hope to mass-produce animals as "bioreactors" to do such things as produce milk with therapeutic proteins to treat human diseases and to create organs for transplantation into humans, Elizabeth Pennisi and Gretchen Vogel report in the following selection. However, although experts have successfully cloned cows, goats, pigs, and mice, numerous problems have riddled animal cloning. Scientists point out that cloning is unpredictable, the process beset with problems at every step, including failed pregnancies, early deaths, and developmental abnormalities. According to Pennisi and Vogel, because of these serious drawbacks, most researchers agree that it is unacceptable to pursue the reproductive cloning of humans. Many scientists also believe that therapeutic cloning, in which stem cells are extracted from clonal embryos and used in treating diseases, may be a long way off. Elizabeth Pennisi is a staff writer for *Science* magazine, an international weekly publication focusing on scientific research. Pennisi's writing specialties are in molecular and cell biology, genetics, and plant sciences. Gretchen Vogel is a writer and European correspondent for *Science* in Berlin, Germany.

Cumulina. Cupid. Peter. Webster. Diana. Dotcom. Dolly. Once the realm of science fiction, cloned animals are now becoming almost commonplace. In the past 4 years, cows, mice, goats, and pigs have joined sheep in an expanding menagerie of cloned mammals. Just around the corner, to judge from the press releases and headlines, looms the next brave new world of biotechnology, with herds of identical cattle, sheep, and goats producing bucketfuls of drugs in their milk, pigs designed to grow spare parts for humans, primates and other animals custom-cloned to study human diseases, and even replacement parts cloned from a patient's own cells. Indeed, given the seemingly endless string of birth announcements, a logical question is, "Will humans be next?"

Not likely, say numerous experts in the field. What the press accounts often fail to convey is that behind every success lie hundreds of failures—some so daunting that many would-be cloners have put efforts to create live animals on hold and are going back to the lab to study why cloning sometimes works but far more often fails. Despite years of effort, "we're in the same bind that we've always been in. A majority of [would-be clones] do not make it to term," says Robert Wall of the U.S. Department of Agriculture (USDA) in Beltsville, Maryland. "We have no explanation; it's more art than science," adds Jean-Paul Renard of the National Institute of Agricultural Research (INRA) in Jouy en Josas, France.

Indeed, even Ian Wilmut, the Scottish researcher who brought the world Dolly, hasn't cloned another animal in years; instead, he is trying to find out what makes cloning by nuclear transfer possible by studying how genes are reprogrammed. Although Wilmut, who works out of the Roslin Institute near Edinburgh, isn't throwing in the towel, he says enormous hurdles must be overcome before cloning becomes practical, much less profitable. First and foremost is the problem of efficiency, which remains at a less-than-impressive 2%; out of some 100 attempts to clone an animal, typically just two or three live offspring result. Even when an embryo does successfully implant in the womb, pregnancies often end in miscarriage. A significant fraction of the animals that are born die shortly after birth. And some of those that survive have serious developmental abnormalities, suggesting that something in the recipe is fundamentally wrong.

What's more, cloning, however arduous, is just the first step. If the goal is to create "bioreactors" that produce therapeutic proteins in milk, or pig pancreases the human body will not reject, then cloners need to insert foreign genes into the genome in exactly the right place—a process that has so far defied most efforts. "The issues are back in the laboratory rather than in the barnyard waiting for something to gestate," says Wall. Cloning veteran Jim Robl of the University of Massachusetts (UMass), Amherst, agrees. Since he and his colleagues at the biotech company Advanced Cell Technology (ACT) in Worcester, Massachusetts, cloned six transgenic calves, they have focused less on producing live offspring than on questions such as whether cloned animals are genetically older or younger than normal.

To increase their odds of producing healthy clones, researchers are now probing fundamental questions of cell biology, such as what kinds of cells make the best donors or what environments are most conducive to the earliest stages of development. They are trying to figure out whether there is something inherently flawed in "asexual" reproduction in mammals—in other words, do we really need two parents? Researchers know that genetic competition between sperm and egg helps to modulate imprinting, a process that selectively silences certain genes early in development. That process may go awry in cloning, accounting for some of the developmental abnormalities.

Or does some problem lie in the "in vitro" component? Toward that end, teams are sorting out more mundane questions of how, exactly, to culture the growing embryo in the lab, and what concoction of hormones is necessary to ensure adequate development.

In all species, the basic hurdles are the same, but the details differ sufficiently that each species has gotten sidetracked at different points along the way to becoming a commercially or medically useful clone. Yet in the highly competitive world of animal cloning, researchers are loath—or sometimes forbidden—to share their tricks of the trade. Added to the normal passions, jealousies and simple desires for credit that plague most high-profile research is the fact that much cloning research is done with corporate sponsorship—and with corporate requirements of secrecy. "Breakthroughs" are often announced long before the technical details are published in journals, making it hard for researchers to verify or extend the results. Even attracting scientists to a recent closed-door conference at the Banbury Center at Cold Spring Harbor Laboratory in New York required a "major diplomatic effort," says molecular biologist Norton Zinder of The Rockefeller University in New York City, who helped organize the meeting to try to promote better communication among the cloners.

Dolly and Friends

With new clones being announced almost monthly, it is easy to forget just how mind-boggling the process really is. Take a single adult cell, whose fate is supposedly sealed, and send it back in time, so to speak, unsealing the genetic instructions contained in that cell's nucleus. Then ask that nucleus, once inserted into another cell, to set that cell on a course of replication and differentiation to produce a whole new animal—one that is a veritable carbon copy of the adult from which that cell came. There's no true biological "mother" or "father" involved.

Through the past half-century, researchers had dabbled with this bold endeavor, transferring nuclei from a variety of cell types and sources into cells whose own genetic material had been removed. Sometimes the nuclear transfer experiments seemed to work. In cows, for example, when nuclei from relatively undifferentiated embryonic cells were put into ripened eggs ready for fertilization, offspring could result. But cloning from an adult or even a fetal cell, which would have begun to differentiate, seemed impossible.

To pull off their paradigm-altering experiment, Wilmut and Keith Campbell at the Roslin Institute spent years painstakingly manipulating both the donor cells and the receiving eggs, developing the finesse to make nuclear transfer work with differentiated cells. Ultimately, they teamed up with PPL [Pharmaceutical Proteins, Ltd.] Therapeutics of Midlothian, Scotland, which wanted to make herds of identical sheep that carried a human gene for a therapeutic protein.

The Roslin group thought they might succeed where others had failed if they could synchronize the cell-division cycle of donor cells with that of the egg. They did this by depriving the donor cells of nutrients, which caused them to shut down most genetic activity. At the same time, they worked out better ways of removing the DNA from the ripened oocytes and, after fusing the oocyte with the donor cell, of triggering cell division as if the egg had been fertilized.

Wilmut, Campbell, and their Roslin colleagues first tried their idea out with embryonic cells, which were presumed to be more malleable. Their success in cloning two lambs in 1996 prompted them to try the same approach using older cells—eventually, cells that they had cultured from an adult ewe's udder. That the experiment worked with the mammary cells was just short of miraculous: Dolly was the product of 434 attempts at nuclear transfer, all but one of which went bad. But that one lamb, reported in February 1997, was enough to set off a worldwide tizzy. Not only did Wilmut's work demonstrate, for the first time, that specialized cells might be reprogrammed and revert to the time when they could become any and all cell types, but it also implied that, if the process worked in sheep, then humans might be just around the corner.

Several colleagues were skeptical, suggesting that perhaps Dolly was the product of a rogue fetal or undifferentiated stem cell and not a true mammary gland cell. But others were in awe and rushed off to try to replicate the results in a variety of animals, from mice to cows. Within a year, companies like ACT and the newly formed Infigen in DeForest, Wisconsin, made public their own successes with cattle, and cloning became a household word.

Meanwhile, Campbell moved over to PPL and set out to take the next step: putting foreign genes into the donor cell's DNA. It didn't take much work to add new DNA to the cultured fetal cells, select those that took in the new genes, and fuse them with enucleated eggs. It worked, as evidenced by the birth of Polly and five other lambs bearing the gene for human factor IX, published in December 1997. Yet even though Polly expressed the new gene— that is, made the protein encoded by the gene—she was just an interim success. Campbell had inserted the gene into a random location in the donor cell's genome, which meant he had little control over how active it would be. To guarantee high production of factor IX protein in sheep's milk—essential if the research was ever to yield a commercially viable bioreactor—Campbell needed to target the gene to a specific spot in the genome. But there was a problem. Until then, gene targeting had only worked in mice and, in one experiment, in human connective tissue cells—and decidedly not in livestock.

That would soon change. In 1997 David Ayares, a molecular biologist at a pharmaceutical company, joined PPL with one goal in mind: gene targeting. The big problem, he, PPL's Alex Kind, and their colleagues quickly surmised, was finding a way to insert the gene before the donor cells got too old. Most cells can divide only a limited number of times in culture, and gene targeting requires several steps in which DNA is inserted and the few select cells that incorporate it into their chromosomes are allowed to multiply until there are enough for the next modification. "By the time you go through and get a large enough population, it's really pushing the limits [of the cells' ability to divide]," says Robl of UMass.

PPL researchers on both sides of the Atlantic set out to improve the efficiency of each step. They also concocted different brews of growth factors that kept the cells healthier through more cell divisions.

In August 1999, at a meeting on transgenic animals, Ayares showed the fruits of this labor: slides of Cupid and Diana, two sheep clones, one containing a marker gene as a control, and one containing both the marker and a gene for alpha-1 antitrypsin, a potentially therapeutic human protein. Both genes were "knocked" into the sheep's genome in much the way gene targeting is done in mice—in other words, they were precisely inserted into the correct spot.

Ayares's announcement was "the most earthshaking news of the year," recalls Wall of USDA, as gene targeting in livestock no longer seemed to be an insurmountable problem. The work has not yet been published, however, and no one has yet repeated the results. Even so, Robl is confident that others will soon succeed. As for PPL, Ayares says the team has pulled off the same feat in cow and pig cells—and that it's just a matter of time before PPL will turn those cells into clones.

Cows: 200 and Counting

Wilmut's success in part grew out of the failures of nuclear transfer in cattle. Wilmut and Campbell built on a technique that researchers had been using for years to "clone" cows from very early embryonic cells. Indeed, in the 1980s, a Texas ag-biotech start-up named Granada had built its business plan on cloning fast-growing cattle this way. At the time, recalls Ken Bondioli, now with Alexion Pharmaceuticals in New Haven, Connecticut, nuclear transfer in cattle had become routine. Although the efficiency was low, "we produced hundreds of calves," he says. But there was a catch. More than the usual number of pregnancies were proving problematic: Deliveries were difficult, and many calves died just before or after birth. "It took us some time to recognize this as a problem having to do with nuclear transfer," Bondioli explains. Eventually, their data led Granada scientists and others to characterize what they called "the large calf syndrome," the cause of which remains a mystery.

Unable to overcome the problem, the company shut down by 1991. Not until Wilmut and the Roslin group produced Dolly and ACT and Infigen had cloned calves with genes randomly inserted into their genomes did nuclear transfer in cattle seem attractive again, says Bondioli. Very quickly, cloning successes in Japan and New Zealand showed that nuclear transfer of fetal or even adult cells worked just fine in cows—or so it first appeared. But as the worldwide total of cloned cattle approaches 300, researchers are finding that they, too, are haunted by Granada's ghost: About a quarter and sometimes more of the calves that survive to birth are bigger than normal, and many of the frustrating spontaneous abortions involve fetuses that are unusually large. Even the normal-sized newborns frequently have lungs like those of premature babies. Others seem to have blood potassium levels high enough that "the calf should be dead," says Michael Bishop of Infigen. Adds Randy Prather of the University of Missouri, Columbia, "Just because you've got offspring doesn't mean they are normal."

In trying to sort out what goes awry, researchers are focusing on two areas. One is imprinting, the critical but poorly understood process by which the protein signal is determined by whether a certain gene came from the mother or the father. Developmental abnormalities result when one copy of the imprinted gene turns on or shuts down inappropriately—a likely prospect in clones, as the whole genome comes from one donor cell rather than the typical two.

Another possibility is that problems arise because of the way the egg is handled before implantation—for instance, if the brew of hormones is not quite right, or if the jostling, poking, and prodding damages the egg in some imperceptible ways. One hint is that cattle conceived in test tubes tend to have some of these same abnormalities. Robl of UMass wonders if they are stymied by both problems: mishandling of the egg and a lack of reprogramming of the donor nucleus. "What we still have is a black box," he admits.

He and others are now systematically evaluating each step. For instance, Bishop and his colleagues at Infigen are keeping a comprehensive database of cell lines, the specific techniques used during nuclear transfer, care and feeding of the surrogate mother, and in utero growth rates in an effort to try to predict which calves will have higher risks of problems. Already they have noticed a correlation between a certain cell line and unusual fluid buildup around the placenta. The company is also using microarrays [snippets of DNA used for measurements] to assess which genes are active in a given cell to see if they can discover a connection between certain genes and cloning success. "In 10 years, I'm sure we'll look back and see how archaic we are," Bishop predicts.

FURTHER CLONING DEVELOPMENTS

The Not-So-Impossible Pig

[In] March [2000] at the Banbury Center meeting, Prather likely experienced one of the worst moments in his career. After almost three frustrating years of trying to clone pigs, he was close to calling that goal unachievable. Hiroshi Nagashima of Meiji University in Tokyo, another speaker that afternoon, had a similar tale of woe. But just before their session was to begin, Alan Colman of PPL Therapeutics made a startling announcement: "I hate to have to say this, given what's coming up, but we've got pigs." A press release about those five piglets, which included a randomly inserted transgene, made the headlines a few days later, although a peer-reviewed scientific paper has yet to be published—convincing most would-be pig cloners that their long-sought goal is now in reach.

Pigs are one of the hottest commodities in cloning, as many scientists believe they are the key to xenotransplantation. Because of their size, pig organs are considered most likely to be compatible with humans and could thus satisfy the unmet need for replacement organs, such as hearts or pancreases. Moreover, some researchers think pig tissue, transplanted, say, in the brain, might be a source for much-needed chemicals that the human body fails to make, such as dopamine, whose loss contributes to Parkinson's disease. So it was no surprise that several animal scientists redirected their research toward cloning pigs once Dolly burst on the scene.

The problem is that until 1998, few scientists had tried to work with immature pig eggs or to grow pig embryos in the lab. Pigs differ from cows and sheep in that they are born in litters, and unless there are at least four viable fetuses in the womb, the pregnancy fails. That means that a day's work has to yield at least several viable embryos if the cloning experiment is to have any chance of success.

Ayares says PPL spent more than a year trying to clone pigs with the techniques the company and Wilmut had used for sheep. Each attempt failed. Pig embryos proved too fragile, and the cells often broke apart during nuclear transfer or handling. In those rare instances when the researchers were able to add the donor nucleus to the egg and then activate development, the embryos never made it to the blastocyst stage. Then in 1998, company scientists junked that approach and hit upon an entirely new—and apparently successful—one that Ayares will not discuss until the scientific paper comes out—other than to say that he is confident that PPL can clone more pigs when they want to. Next time around, he says, the company plans to genetically modify the donor cells to make pig organs more acceptable to the human immune system—a key step toward making xenotransplantation a reality.

Prather, however, is withholding judgment until he sees more piglets. "I think they got lucky," he says of PPL.

Filling Out the Barnyard

Cloners are testing the waters on other livestock, with mixed success. Goats, it seems, are easy. In 1999, two companies reported . . . that they had successfully cloned goats. And [in 2000], Nexia Biotechnologies Inc. of Montreal announced the birth of two goats, Webster and Peter, that carry a gene from arachnids that codes for the spider silk protein. This spring, says a company spokesperson, Nexia mated the two bucks with normal females; by year's end they expect the female offspring will be churning out milk chockfull of spider silk protein. The company plans to extract the protein and spin it into light, high-strength fibers for use as sutures or in bulletproof vests or automotive and aerospace components.

On the other hand, despite their natural fecundity, rabbits have so far defied efforts to produce them in the lab. "We can get a lot of cloned embryos," says Renard of the National Institute of Agricultural Research in France, but all pregnancies abort after transfer to a surrogate mother. He suspects that the problem may be in the earliest cell divisions in the embryo. Renard and his colleagues will keep trying, however, as transgenic rabbits produced by cloning could be valuable tools for studying cardiovascular disease.

In terms of bioreactors, it would be tough to beat the chicken—or more precisely, its egg, says USDA's Wall—which is why several companies are now trying to clone chickens. A typical egg costs about 2 cents to make, and if cloners can insert a therapeutic gene and get it to express in the egg white, commercial technology already exists for separating the whites from the yolks. But the eggs themselves present cloners with a distinct problem: their huge size, says Leandro Christmann, a reproductive biologist at AviGenics in Athens, Georgia. And size does matter. To remove DNA from mammalian eggs with diameters of roughly 100 micrometers, you simply put the fairly transparent egg under a microscope and suck out the DNA with a pipette. But in chickens, the egg yolk is far too big and opaque. Just figuring out how to "see" the chicken egg nucleus has been a challenge, Christmann says. Even so, the start-up says it has made progress in developing or adapting the technology to take that egg through all the necessary steps, prompting Avi-Genics president Carl Marhaver to predict that within a year, "we will be able to produce the first cloned bird."

If Primates, Then Humans?

Perhaps no area of cloning research evokes more curiosity than primates. Although researchers aren't attempting primates as a dry run for humans—their goal is to create identical animals to study such diseases as hepatitis—their progress is likely to shed light on when it might be technically possible to clone people.

Those worried that some crazed scientist might ignore the ethical and legal sanctions against human cloning experiments and plow right ahead can rest assured: It won't be easy, at least according to Tanja Dominko, a reproductive physiologist at Oregon Health Sciences University in Portland.

When she arrived in Oregon in 1997, prospects looked fairly bright. Oregon's Don Wolf had just succeeded in using nuclear transfer to produce two monkeys, Neti (Nuclear Embryo Transfer Individual) and Ditto, a year earlier. As Granada had done with cows in the 1980s, Wolf had used nuclei from embryonic cells—far easier to work with than adult cells. But 300 attempts and no pregnancies later, the picture "is not as rosy," Dominko says. Cloning primates "is not just around the corner." Neither she, working with Oregon's Gerald Schatten, nor Wolf's team working one floor below, has been able to replicate Wolf's early success.

Once Dominko realized what she was up against, she tried to determine whether the problem was with nuclear transfer or the in vitro procedures. She attempted "mock" nuclear transfers, in which she and her colleagues went through the cloning procedure but didn't actually replace the egg's own DNA. Instead, they fertilized the egg in vitro after poking and prodding it the way they would have for a true nuclear transfer experiment, and then placed it in a female for gestation.

Those attempts didn't work well, suggesting to Dominko that the in vitro procedures were the problem. She then made "egg-friendly" improvements such as using sperm extract instead of harsher chemicals to prompt the egg to divide, which helped the subsequent nuclear transfer experiments. The team produced roughly 45 embryos by nuclear transfer this way, but none successfully implanted in a surrogate female monkey's womb.

Then Dominko, Schatten, and colleagues began looking at the transferred nucleus itself. Under a light microscope, the embryo's expanding cluster of cells, known as the blastocyst, looked just like those seen after successful in vitro fertilization. But a closer look at these dividing cells, with confocal microscopy, revealed "a whole gallery of horrors," says Schatten. The new nucleus seemed completely out of sync with the egg. Even the first cell division had gone awry, as the chromosomes didn't seem to have copied and separated as they should have. By the eight-cell stage, some cells had too much DNA, while a few seemed to have none at all.

Dominko and Schatten then took a closer look at the spindle, and in particular at the centrosomes, which help organize and guide the movement of DNA during cell division, making sure that each cell gets the right complement of chromosomes. In primates, they found, the incoming nucleus tends to leave behind one or both of its centrosomes. "The embryos we were making probably never had a chance," says Dominko. Given these results, which are still unpublished, Schatten and Dominko have all but given up on cloning by nuclear transfer until they develop a better understanding of these abnormalities. Instead, they have turned to embryo splitting, in which the early embryo is divided in two and gives rise to identical twins, as a means of generating like animals useful for research. Dominko and Schatten don't know why primates are different from cows, but they are convinced that attempts to clone humans would run up against the same biological roadblocks.

Even Wolf on the floor below is now looking at embryo splitting, but he has not abandoned nuclear transfer. He's tried nuclear transfer with some 100 embryos, none of which has established a pregnancy. Indeed, his studies have revealed another source of failures: Embryos don't develop at the same rate in a lab dish as they do in the womb. It's important to keep trying, he argues, as clinical studies often require more than the two identical animals that can be produced by embryo splitting.

Litters and Litters of Mice

Despite efforts by numerous labs to clone mice, this laboratory staple has proved remarkably elusive. Indeed, until recently, only one person in the world had been able to pull it off: Teruhiko Wakayama, who originally reported success with Ryuzo Yanagimachi at the University of Hawaii, Honolulu, in July 1998.

Even in Wakayama's skilled—and some say "magic"—hands, cloning is unpredictable and enigmatic. Fatal problems can crop up at every step. Even the temperature of the lab can make a difference: Slightly too hot or cold, and the technique won't work as well. When Wakayama first moved from Honolulu to Rockefeller University in November 1999, nothing seemed to go right. For the first few months, few of the embryos survived, says neuroscientist Peter Mombaerts, who helped lure Wakayama and his colleague Tony Perry to Rockefeller. One contributing factor, the team suspects, was that their brand-new incubator was producing toxic gases and killing the embryos.

Even without toxic incubators, other labs remained frustrated in their efforts to duplicate Wakayama's work, prompting some disbelief. Wakayama and Yanagimachi's procedure involves injecting the nucleus into the enucleated oocyte instead of fusing the entire donor cell with an electrical charge, and this feat requires the steady hand of a surgeon. "It requires very miniature handwork, and it has to go reasonably fast," explains Mombaerts. Adds Perry: Wakayama "certainly has the magic touch."

But now others working with Yanagimachi and at least three new labs are claiming to have the magic touch as well. Atsuo Ogura of the National Institute of Infectious Diseases in Tokyo and his colleagues have cloned mice with the Honolulu technique. . . . Renard says his group in France has also produced a few litters, with several more pregnancies under way. And after tutoring from the Hawaii team, postdoc William Rideout and graduate student Kevin Eggan in Rudolf Jaenisch's laboratory at the Massachusetts Institute of Technology have also successfully repeated the technique. Most agree that Wakayama has a knack for the injections, but "the technique is transferable," Jaenisch says.

That is good news for the cloning field as a whole, Jaenisch says. Scientists are eager to use cloned mice as a powerful lab tool—not the least to study cloning itself. Because mice are small and reproduce quickly, and because scientists know so much about their genetics and their development, researchers say the mouse offers the best hope for answering many of the questions that plague efforts to clone other species.

One of those other species, of course, is humans. For now, the serious obstacles to cloning every species, especially other primates, suggest that human cloning—even so-called therapeutic cloning to produce cell lines to be used in treating disease—may be a long way off. As for reproductive cloning, or actually creating a living replica, "it would be criminal at this stage in our abilities," says Zinder of Rockefeller. Most researchers concur. The U.S. National Bioethics Advisory Commission issued a report in 1997 saying that human reproductive cloning would be unethical for a variety of reasons, and the commission's chair, Harold Shapiro of Princeton University, says it is still "clinically and scientifically premature to produce human infants." Even if the technique were safe, he adds, it would be unethical to proceed without a clearer public consensus. Given the huge scientific unknowns, there should be ample time for sorting out whether human cloning would ever be acceptable should it, too, yield to the magician's touch.

Handmade Cloning: A New Technique

By Sylvia Pagán Westphal

A new technique called handmade cloning could help roboticize the process of cloning farm animals, according to Sylvia Pagán Westphal in the following selection. In addition to mass-producing the best milk- and meat-producing animals, scientists could use handmade cloning to reproduce endangered species in places like South Africa. Compared to the old technique, handmade cloning requires less expensive equipment and less expertise to perform. The old way of cloning involved an expensive machine called a micromanipulator that allowed a skilled technician to remove a cell's nucleus using a very fine needle. In this new technique, egg cells are split in half and the halves containing the nucleus are discarded. A cell from an adult animal is then fused with two of these empty egg cells by zapping them with an electric current. The technique has initially yielded higher survival rates in cloned cattle. Sylvia Pagán Westphal is a writer and Boston correspondent for the *New Scientist* magazine, a leading science and technology weekly.

Handmade cloning, a new way to create genetically identical copies of animals, is not only cheaper and simpler than existing methods, but appears to work better too.

"It's so much simpler than anything we are doing today, it's dramatic," says Michael Bishop, ex-president of Infigen, a cattle-cloning company in Wisconsin. "It's a huge step towards roboticising the whole process."

The technique could speed up the introduction of cloning in

Sylvia Pagán Westphal, "So Simple, Almost Anyone Can Do It," *New Scientist*, vol. 175, August 2002, p. 16. Copyright © 2002 by Reed Business Information, Ltd. Reproduced by permission.

farming, where the aim is to clone the best milk or meat-producing animals. And conservationists in South Africa could soon use it to clone endangered species.

The technique was developed by Gábor Vajta at the Danish Institute of Agricultural Sciences in Tjele together with Ian Lewis, programme leader for the Cooperative Research Centre for Innovative Dairy Products in Australia. Details of the method will soon be published.

At the moment, the key instrument in cloning is the "micromanipulator", an expensive machine that allows a skilled technician to grab an egg cell under the microscope, insert a very fine needle to suck out its nucleus, and then use another needle to transfer a nucleus from the animal to be cloned. An alternative is to fuse the empty egg with a cell from the animal. Either way, it's a tricky and time-consuming process.

How the New Technique Works

In the new technique, egg cells are split in half under a microscope using a very thin blade. The halves quickly seal up. A dye is used to identify the halves containing the nucleus, which are then discarded, leaving only empty "cytoplasts". To create a cloned embryo, a cell from an adult animal is fused first with one cytoplast, then another, by briefly zapping them with an electric current.

Half of the cow embryos created this way survive long enough to form balls of cells called blastocysts, ready to be implanted in the womb. That success rate is at least as good as current standards.

But the big advantage this method has over normal cloning is that it you can use relatively cheap equipment, and personnel can be trained very quickly. It should be a boon to researchers on tight budgets, Vajta says. It should also be far easier to automate. Some companies are already trying to develop chips that mass-produce cloned embryos.

A healthy-looking calf created by handmade cloning has already been born in Australia, and another is expected [in August 2002]. In preliminary tests, the Danish researchers implanted seven cow blastocysts, resulting in six pregnancies. After 150 days—the threshold after which cattle pregnancies usually carry to term—three are still pregnant. By comparison, a recent paper suggests an average of only 25 per cent of cows implanted with embryos cloned the standard way are pregnant after 30 days.

It's too early to tell whether the animals created using this new technique will be healthier than those from normal cloning, which often fail to carry to term or have birth defects. But Vajta thinks the reduced manipulation times and the use of two cytoplasts should yield better results. In the normal method, up to a third of the egg's cytoplasm can be lost when the nucleus is removed, whereas fusion with two cytoplasts produces embryos with the same volume as the original egg.

Setting Up a Cheap Lab

[In July 2002], Paul Bartels and his team from the Endangered Wildlife Trust in Johannesburg tried out the method under field conditions. A Bunsen burner on a lab bench served as the sterile working area, and the most expensive piece of equipment was the electrofusion machine, still relatively cheap at $3500. "One can set up a lab very cheaply. You can imagine doing this in a trailer," he says.

The team fused cow cytoplasts with adult cells from endangered species such as the darted buffalo, the bontebok (a kind of antelope), the giant eland and the black impala. "We were very surprised at the health of the embryos. They looked so good," Bartels says.

The team also put five cloned cow embryos into three cows. If this results in healthy calves, they will consider cloning endangered species using closely related common species both as a source of eggs and as surrogate mothers.

Another advantage of the method is that it may bypass existing cloning patents. One worry, however, is that the method's simplicity will make it easier for maverick doctors to attempt human reproductive cloning. But there is one deterrent. They will need twice as many eggs as normal—and human eggs are in very short supply.

The Failure to Clone Monkeys Throws Doubts on Human Cloning

By Gretchen Vogel

In the following selection Gretchen Vogel reports that researchers are having problems trying to clone rhesus monkeys. A research team found that the usual process of somatic cell nuclear transfer, where the nucleus from one cell is extracted and then injected into an egg whose nucleus has been removed, does not work with monkeys as it has with sheep, cattle, and other animals. According to Vogel, researchers say that primate eggs are biologically different from the eggs of other mammals and that in the cloning process, the monkey embryos are robbed of key proteins. The loss of these proteins disturbs the cell's chromosome count and ability to divide properly. Studies suggest that even if this obstacle to primate cloning is overcome, further problems exist for clones of all species, including developmental problems, which often result in premature death. One scientist says that nature appears to have put its own limits on cloning, and Vogel concludes that these problems put human cloning out of reach for the present. Gretchen Vogel is a writer for *Science*, a weekly magazine on scientific topics.

While governments around the world debate how to prevent human reproductive cloning, it seems that nature has put a few hurdles of its own in the way. A team reports that in rhesus monkeys, cloning robs an embryo of key proteins that allow a cell to divvy up chromosomes and divide properly. Unpublished data from this and other groups suggest that the same problem may also thwart attempts to clone humans.

There are potential ways around the newfound obstacle, but for now, groups that made controversial claims that they would use the techniques that produced Dolly the sheep to create human babies are unlikely to succeed.

It is almost as if someone "drew a sharp line between old-world primates—including people—and other animals, saying, 'I'll let you clone cattle, mice, sheep, even rabbits and cats, but monkeys and humans require something more,'" says Gerald Schatten of the University of Pittsburgh School of Medicine, a leader of the rhesus monkey study.

Schatten and his colleagues have tried hundreds of times to clone monkeys, only to fail. Indeed, although several groups have attempted it, no one has yet produced a monkey through somatic cell nuclear transfer, the process by which a nucleus from one cell is extracted and injected into an egg whose own nucleus has been removed. "The failure to clone any primate has so far been startling" says Rudolf Jaenisch of the Massachusetts Institute of Technology in Cambridge, who studies cloning in mice.

The Wrong Number of Chromosomes

The scientists had suspected for several years that something was disturbing cell division in cloned embryos. The embryos seemed normal at their earliest stages, but none developed into a pregnancy when implanted. When the researchers looked more closely, they realized why: Many of the cells in a given embryo had the wrong number of chromosomes. Some had just a few, whereas others had twice as many as they should. Although embryos can survive for a few cell divisions with such defects, soon the developmental program becomes hopelessly derailed.

To find out what was interfering with proper cell division, the team fluorescently labeled the cell-division machinery. The cells' mitotic spindles, which guide chromosomes to the right place during cell division, were completely disorganized. And two proteins that help organize the spindles, called NuMA and HSET, were missing.

A look at unfertilized rhesus oocytes explained why. The team found that the spindle proteins are concentrated near the chromosomes of unfertilized egg cells—the same chromosomes that are removed during the first step of nuclear transfer. In most other mammals, Schatten says, the proteins are scattered throughout the egg, and removing the egg's chromosomes seems to leave enough of the key proteins behind for cell division to proceed.

The work "explains why no one has yet succeed[ed] in achieving normally developing embryos from human nuclear transfer," says Roger Pedersen of the University of Cambridge, U.K., who attempted human nuclear transfer experiments at his previous laboratory at the University of California, San Francisco. "Primate eggs are biologically different." Schatten says preliminary data suggest the proteins are also concentrated near the nuclear material in unfertilized human eggs.

A cloning lab might surmount the hurdle, says Schatten, by reversing the order of the traditional nuclear transfer procedure: First add an extra nucleus, then activate cell division, and finally remove the egg's DNA. The find "will make people think differently about the optimum sequence of nuclear transfer procedures," says Ian Wilmut of the Roslin Institute in Midlothian, Scotland, a leader of the team that cloned Dolly [the Sheep in 1996].

Even if scientists could overcome the obstacles, however, another study suggests that further developmental problems threaten clones of all species. Jaenisch and his colleagues report in . . . *Development* that genes important to early development frequently fail to turn on in mouse embryos cloned from adult cells. That failure helps explain the low survival rate of such embryos, Jaenisch says. But he notes that the team's work—which examined the expression of just 11 genes—is only the tip of the iceberg. In other experiments, the researchers have found that even apparently healthy cloned mice show abnormal levels of gene expression. "There may be no normal clones," Jaenisch says.

Although revising the nuclear transfer procedure might help solve the cell-division problem, it is harder to imagine a solution for the faulty gene regulation that Jaenisch and his colleagues see. "We're looking at a more fundamental problem," he says.

The biological roadblocks would seem to be good news for those worried about the ethical implications of human cloning, says Schatten. "This reinforces the fact that the charlatans who claim to have cloned humans have never understood enough cell or developmental biology" to succeed, he says. The debate will go on, but nature already seems to have imposed its own limits on cloning.

Cloning Endangered Species

By Tamara Levine

In this selection Tamara Levine reports that the first successful endangered species was cloned in January 2001. A gaur, a type of large East Asian ox, was successfully cloned but died two days after its birth because of common dysentery not related to the cloning procedure. Although some researchers heralded the event as a positive step in endangered species preservation, ethicists were concerned that it would bring scientists closer to cloning humans. Furthermore, Levine reports, conservationists have many doubts about the benefits of cloning endangered species, including the high cost of the procedure, and the fact that it does not address the more important questions of why the species are becoming extinct in the first place. Tamara Levine attends the University of Waterloo, Ontario, Canada, where she is a student in environment and resource studies.

The birth of Noah, a clone of a rare gaur, marked the first successful cloning of an endangered species by means of a cross-species nuclear transfer technique. While the researchers trumpet the technique as a valuable new way of saving endangered species, some ethicists worry that the project brings science a step closer to cloning humans. Even species preservation advocates are wondering whether the development is positive.

In January [2001], Advanced Cell Technology (ACT), a Massachusetts-based biotech firm, announced that a cow impregnated with an egg containing genetic material from a gaur, a type of large ox native to east Asia, had given birth to a male gaur calf, named Noah. Two days later the calf died of common dysentery unrelated to the cloning process.

ACT head Michael West says the successful birth represents a new way of preserving diminishing genetic diversity in endangered animal populations. But critics note that the company's ultimate objective is to use the technology in human applications. West has stated, "The goal of this research is to use the cross-species cell transfer technology to reprogram human cells for medical purposes." In the meantime, it is being used to preserve genetic diversity in endangered animal populations.

ACT is making the cloning technique available to any organization that requests it (the company will provide its services and technical assistance for a fee). Researchers plan to clone a wide range of species—including the newly extinct Spanish mountain goat, the burcardo. The last burcardo died [in 2001], and the remains were frozen in anticipation of cloning it.

Other cloning plans focus on the endangered Chinese panda, the endangered Tasmanian tiger, the regionally extinct Barbary lion, the extinct mammoth and the regionally extinct Indian cheetah.

"We don't see this as a profit centre for the company," says West. "Our thought is simply that the human species has casually used technology to despoil the planet; the least we can do is to use the technologies we work with every day to make a small contribution to save innocent and endangered species."

Cloning Is Not an Answer to the Extinction Crisis

The techniques involved in cloning are expensive: Noah's creation cost ACT (US)$200,000. Some conservationists argue that this type of investment is ill-directed because it does not address the fundamental causes of species extinction. "It's not an answer to the extinction crisis," says George Amato, director of the Science Resource Centre at the Bronx Zoo. "It's complicated, it's expensive and it doesn't work all the time."

Noah was the result of 690 attempts to create embryos from the gaur skin cells. Of these, 30 were transferred to cow eggs; eight of the cows became pregnant, and only one carried to term.

There are also problems with the availability of surrogate mothers. Host animals must be genetically compatible and of proper physical dimensions to carry an animal to term. For example, researchers are currently trying to develop a surrogate host for the Chinese panda. Bears in captivity have been examined as potential hosts, but scientists fear fetus rejection by the host and birth complications.

Another problem is that cloned animals will always be the same gender as their parent. This may result in a lack of sexual diversity essential for natural reproduction, which is especially problematic where extinct species are being regenerated. Also, low genetic diversity of the cloned species leaves them susceptible to disease and stress.

There are also questions about interactions between host DNA and the genetic material of the clone. Although the nucleus of the host egg has been removed, the remaining egg does have some genetic material that may influence development and result in impure species.

Conservationists Raise Fundamental Questions

Conservationists raise a more basic question. They doubt that it is worth saving the genetic diversity in a population when there is no habitat to support the species.

"The fundamental question is, why are these populations disappearing in the first place?" says Brewster Kneen, author of *Farmegeddon*, a book critical of agricultural biotechnology. The major reasons for species decline are loss of habitat, human encroachment and excessive hunting, trapping or poisoning.

Already one million US dollars have been invested in preparations for cloning the Indian cheetah. It's money some conservationists think would be better invested in saving existing species and their habitats. No cheetahs are left in India, and the genes for the project will have to be imported from another country.

Conservationists fear that the public will prefer the more glamorous and high profile cloning over breeding and habitat preservation programs that take longer. "The danger is that this could be seen as an alternative," says *Scientific American* editor John Renni.

Bill Holt of the London Zoological Society agrees. "Unless there is a really special reason, and there may be in a small number of cases, we should go for much simpler techniques for conservation," he explains.

Scientists at ACT concede that conservation is still important, but argue that cloning preserves essential biodiversity and does not threaten conservation efforts. "There are hierarchies of priorities, and, in mine, saving habitats of existing, viable species is a higher priority," ACT vice president Robert Lanza told the *Guardian Observer*. "Reconstructing something that's extinct, while having a certain fascination, is less urgent than saving more things from becoming extinct. But I'm not saying this is a waste of time, trivial or wrong."

Yet, many argue that this process is inherently wrong. One of the major controversies centres on the potential application of cloning to human embryos. In 1998, ACT scientist Jose Cibelli used his own leg skin cells, and a cow's egg from which the nucleus had been removed, to develop a human clone. The clone was terminated early in development, but for some ethicists it raised frightening possibilities.

Ethicists fear that the application of this technology to human populations could lead to "genetic elitism," with unequal access to genetic treatments, eugenics movements and compromised rights of individuals and fetuses.

adult stem cell: An undifferentiated cell found in a differentiated tissue in an adult organism that can renew itself and can (with certain limitations) differentiate to yield all the specialized cell types of the tissue from which it originated.

asexual reproduction: Reproduction not initiated by the union of oocyte (egg) and sperm. Reproduction in which all (or virtually all) the genetic material of an offspring comes from a single progenitor.

assisted reproductive technologies (ARTs): Fertility treatments or procedures that involve laboratory handling of gametes (eggs and sperm) or embryos. An example of ARTs is in vitro fertilization.

blastocyst: The very early embryo consisting of about 30–150 cells. The name used for an organism at the blastocyst stage of development.

blastocyst stage: An early stage in the development of embryos, when (in mammals) the embryo is a spherical body comprising an inner cell mass that will become the fetus surrounded by an outer ring of cells that will become part of the placenta.

blastomere: The name for each of the cells of the early two-celled embryo after its first cell division.

cell line: A general term applied to a defined population of cells that has been maintained in culture for an extended period and usually has undergone a spontaneous process, called transformation, that allows the cells to continue dividing (replicating) in culture indefinitely.

chromosomes: Structures inside the nucleus of a cell, made up of long pieces of DNA coated with specialized cell proteins, that are duplicated at each cell division. Chromosomes transmit the genes of the organism from one generation to the next.

clone: 1) An exact genetic replica of a DNA molecule, cell, tissue, organ, or entire plant or animal. 2) An organism that has the same nuclear genome as another organism.

cloned embryo: An embryo arising from the somatic cell nuclear transfer process as contrasted with an embryo arising from the union of an egg and sperm.

cloning: The production of a clone.

culture: The growth of cells, tissues, or embryos in vitro on an artificial nutrient medium in the laboratory.

cytoplasm: The contents of a cell other than the nucleus. Cytoplasm consists of a fluid containing numerous structures, known as organelles, that carry out essential cell functions.

differentiation: In cells, the process of changing from the kind of cell that can develop into any part of the body to a cell of one specific kind.

diploid: Refers to a cell having two sets of chromosomes (in humans, forty-six chromosomes in all).

DNA: A chemical, deoxyribonucleic acid, found primarily in the nucleus of cells (some is also found in the mitochondria). DNA is the genetic material that contains the instructions for making all the structures and materials the body needs to function. Chromosomes and their subunits, genes, are made up (primarily) of DNA.

embryo: 1) The developing organism from the time of fertilization until significant differentiation has occurred, when the organism becomes known as a fetus. 2) An organism in the early stages of development.

embryo cloning (artificial twinning): A medical technique that produces identical twins or triplets. One or more cells are removed from a fertilized embryo and encouraged to develop into one or more duplicate embryos.

embryonic stem (ES) cells: The primitive (undifferentiated) cultured cells from the embryo that have the potential to become a wide variety of specialized cell types (that is, they are pluripotent). These cells by themselves cannot produce the necessary cell types in an organized fashion so as to give rise to a complete organism.

embryo splitting or **twinning:** The separation of an early-stage embryo into two or more embryos with identical genetic makeup, essentially creating identical twins or higher multiples (triplets, quadruplets, etc.).

enucleated egg: An egg cell whose nucleus has been removed or destroyed.

eugenics: An attempt to alter (with the aim of improving) the genetic constitution of future generations.

gamete: A reproductive cell (egg or sperm). Gametes are haploid (contain twenty-three chromosomes).

gene: A functional unit of heredity that is a segment of DNA in a specific site on a chromosome.

genome: The complete genetic material of an organism.

germ cell or germ line cell: A sperm or egg, or a cell that can develop into a sperm or egg; all other body cells are called somatic cells.

haploid: Refers to a cell having only one set of chromosomes (twenty-three chromosomes in all in humans).

in vitro fertilization (IVF): The union of an egg and sperm, where the event takes place outside the body and in an artificial environment (the literal meaning of "in vitro" is "in glass"; for example, in a test tube).

mitochondria: Small energy-producing organelles inside of cells, mitochondria give rise to other mitochondria by copying their small piece of mitochondrial DNA and passing one copy of the DNA along to each of the two resulting mitochondria.

multipotent cell: A cell that can produce several different types of differentiated cells.

nonsomatic cells: These are germ cells (sex cells) such as sperm and egg cells. They are haploid (containing only twenty-three chromosomes).

nuclear transfer: A procedure in which a nucleus from a donor cell is transferred into an enucleated egg or zygote (an egg or zygote from which the nucleus has been removed). The donor nucleus can come from a germ cell or a somatic cell.

nucleus: An organelle, present in almost all types of cells, that contains all of the cell's genes except those in the mitochondria.

oocyte: The developing female reproductive cell (developing egg) produced in the ovaries.

organism: Any living individual animal considered as a whole.

parthenogenesis: A form of nonsexual reproduction in which eggs are subjected to electrical shock or chemical treatment in order to initiate cell division and embryonic development.

pluripotent: A cell that can give rise to many different types of differentiated cells. A pluripotent cell can form nearly all or many kinds of tissue.

precursor cells or **progenitor cells:** In fetal or adult tissues, these are partially differentiated cells that divide and give rise to differentiated cells.

somatic cell: Any cell of a plant or animal other than a reproductive cell or reproductive cell precursor, also called body cells. These are diploid cells (containing two sets of chromosomes, which total the forty-six chromosomes constituting the DNA of that species).

somatic cell nuclear transfer (SCNT): Transfer of the nucleus from a donor somatic cell into an enucleated egg to produce a cloned embryo.

stem cells: Undifferentiated cells found in embryos and in small amounts in adult tissue, which are capable of both perpetuating themselves as stem cells and of undergoing differentiation into one or more specialized types of cells.

telomeres: "Caps" (made of repeated DNA sequences) found at the ends of chromosomes that protect the ends of the chromosomes from degradation. The telomeres on a chromosome shorten with each round of cell replication. Telomere shortening has been suggested to be a "clock" that regulates how many times an individual cell can divide (that is, when the telomeres of the chromosomes in a cell shorten past a particular point, the cell can no longer divide).

totipotent: A cell with an unlimited developmental potential, such as the zygote and the cells of the very early embryo, each of which is capable of giving rise to 1) a complete adult organism and all of its tissues and organs, as well as 2) the fetal portion of the placenta.

transgenic: A term that refers to an organism that contains genes from more than one species, usually as a result of genetic engineering.

undifferentiated: Refers usually to an embryonic or fetal cell, the DNA of which has not been programmed to a specific cell type. These can turn into almost any type of cells, and therefore are totipotent.

xenotransplantation: The transplantation of tissues or organs from one species to another.

zona: The shell that surrounds the oocyte (egg) and young embryo before implantation.

zygote: The diploid cell that results from the fertilization of an egg cell by a sperm cell.

1894

Hans Dreisch isolates blastomeres (cells that develop in the first stage of cell division after the egg is fertilized) from two- and four-cell sea urchin embryos and observes their development into small larvae.

1901

German embryologist Hans Spemann, called "the father of modern embryology," splits a two-cell salamander embryo into two parts, resulting in the development of two complete larvae. This proved that early embryos carry all the genetic information required to create an organism.

1902

Walter Sutton publishes works hypothesizing that chromosomes carry inheritance and that they occur in distinct pairs within a cell's nucleus.

1928

Spemann performs the first successful early nuclear transfer experiments, cloning a salamander.

1938

Spemann publishes the results of his 1928 primitive nuclear transfer experiments involving salamander embryos in the book *Embryonic Development and Induction*. He proposes a "fantastical experiment" to transfer a late-stage embryo cell's nucleus into an egg without a nucleus, the basic method that would eventually be used in cloning.

1944

Oswald Avery finds that a cell's genetic information is carried in DNA.

1952

Robert Briggs and Thomas J. King add the nucleus from an advanced frog embryo cell to a frog egg, cloning northern leopard frogs. Some call this the first cloned animal.

1953

Francis Crick and James Watson, working at Cambridge's Cavendish Laboratory, discover the structure of DNA. The men receive the Nobel Prize for their work.

1962

Oxford University biologist John Gurdon, called "the pioneer of modern cloning," announces that he has cloned South African frogs using the nucleus of fully differentiated adult intestinal cells. This demonstrates that cells' genetic potential does not diminish as the cell becomes specialized. His claim generates heavy debate and interest in cloning.

1963

British biologist J.B.S. Haldane coins the term "clone" in a speech entitled "Biological Possibilities for the Human Species of the Next Ten-Thousand Years." Even though many scientists have described and even completed the cloning process by this time, the term "cloning" has never been used to describe such experiments.

1966

Marshall Niremberg, Heinrich Mathaei, and Severo Ochoa break the genetic code. The cracking of the genetic code opens the door for the explosion of genetic engineering studies and achievements beginning in the late 1970s.

1968

James Shapiera and Johnathan Beckwith announce that they have isolated the first gene. Their discovery is part of a wave of molecular biology discoveries directly following the 1966 cracking of the genetic code. The announcement also increases the public's concern about the growing power of molecular biologists.

1972

Paul Berg of Stanford University combines the DNA of two different organisms, thus creating the first recombinant DNA molecules.

1973

Stanley Cohen and Herbert Boyer create the first recombinant DNA organism using recombinant DNA techniques pioneered by Paul Berg. Also known as gene splicing, this technique allows scientists to manipulate the DNA of an organism, the basis of genetic engineering.

1979

Karl Illmensee and Peter Hoppe claim to have cloned three mice by transplanting the nuclei of mouse embryo cells into mouse eggs. Other scientists are unable to reproduce the results. The validity of the research is still questioned.

1980

In the case *Diamond v. Chakrabarty*, the U.S. Supreme Court rules that a "live, human-made microorganism is patentable material." This paves the way for making cloning research profitable.

1984

Steen Willadsen, a Danish scientist, reports that he has made a genetic copy of a lamb from early sheep embryo cells, a process now called "twinning" (a term sometimes used interchangeably with "cloning"). This is the first verified cloning of a mammal using nuclear transfer. Other scientists eventually used his method to "twin" cattle, pigs, goats, rabbits, and rhesus monkeys. Within two years, Willadsen used differentiated one-week-old embryo cells to clone a cow, proving that the DNA of specialized cells could be returned to its original state.

October 1990

The National Institutes of Health officially launches the Human

Genome Project, a massive international collaborative effort to locate the fifty thousand to one hundred thousand genes and sequence the estimated 3 billion nucleotides of the human genome. By determining the complete genetic sequence, scientists hope to begin correlating human traits with certain genes. With this information, medical researchers have begun to determine the intricacies of human gene function, including the source of genetic disorders and diseases.

1993
Human embryos are first cloned.

1994
Neal First produces genetic copies of calves from embryos. They grow to at least 120 cells.

July 1995
Ian Wilmut replicates First's experiment with differentiated cells from sheep, but he puts the embryo cells into an inactive state before transferring their nuclei to sheep eggs. The eggs develop into normal lambs, named Megan and Morag.

July 5, 1996
Dolly, the first animal (clone) ever to be created from adult cells, is born. Her birth is not announced until February 1997 when Ian Wilmut and Keith Campbell at the Roslin Institute in Scotland report that they have cloned an udder cell of a six-year-old adult sheep. The scientists delayed the announcement in order to get patent rights on the technology they used.

1997
March: Only a week after the Dolly announcement, the Oregon Regional Primate Research Center scientists bring cloning technology closer to humans by twinning two rhesus monkeys from embryos.
June: President Bill Clinton signs a five-year moratorium on the use of federal funds for human cloning research; his National Bioethics Advisory Commission concluded that human cloning would be unsafe and unethical.

July: Wilmut and Campbell, the scientists who created Dolly, create Polly, a Poll Dorset lamb clone, from skin cells grown in a lab and genetically altered to contain a human gene in every cell of its body. The first of its kind, Polly's birth is the first step in the application of cloning technology to the production of a useful product. Wilmut's and Campbell's creation of Polly surprises the scientific community by demonstrating how fast cloning technology is progressing. The cloning of genetically altered farm animals was not expected for another five years.

August: Clinton proposes legislation to ban the cloning of humans for at least five years and a month later, in September, thousands of biologists and physicians sign a voluntary five-year moratorium on human cloning in the United States.

December: Harvard graduate Richard Seed announces that he intends to clone a human before federal laws can effectively prohibit the process. Seed's announcement rekindles the raging ethical debate on human cloning sparked by the creation of Dolly.

1998

January: Nineteen European nations sign a ban on human cloning. The Food and Drug Administration (FDA) in the United States announces that it has authority over human cloning, thus making it a violation of federal law for anyone to try the procedure without FDA approval.

July: Researchers at the University of Hawaii announce that they have cloned fifty mice from adult cells. Some of the mice are clones of clones, created by using a technique different than that used to produce Dolly the sheep.

January 2000

Britain becomes the first country to grant a patent for cloned early-stage human embryos. Geron Corporation, which received the patent, said it has no intention of creating cloned humans. More animals are cloned around the world.

July 2001

The U.S. House of Representatives approves legislation that would make it a federal crime to clone people to produce children or to create embryos for medical research.

2002

February: A House of Lords committee gives Britain's scientists a green light to pioneer the cloning of human embryos for research and set up the world's first embryo cell bank.

November: Dr. Severino Antinori, an Italian fertility expert, says he expects a patient to deliver a healthy cloned baby in January 2003 and that two other women are carrying cloned embryos.

December: Clonaid, a company associated with the Raelian movement which believes mankind was created by aliens, announces it has created the first human clone, a baby girl. In January 2003 the independent scientist and journalist brought in by Clonaid to verify its successful cloning denounces their claim as a possible "elaborate hoax."

February 2003

Six-year-old Dolly the sheep is given a lethal injection after signs of progressive lung disease are discovered.

FOR FURTHER RESEARCH

Books

Lori B. Andrews, *The Clone Age: Adventures in the New World of Reproductive Technology.* New York: Henry Holt, 1999.

Andrea L. Bonnicksen, *Crafting a Cloning Policy.* Washington, DC: Georgetown University Press, 2002.

Michael Brannigan, ed., *Ethical Issues in Human Cloning: Cross-Disciplinary Perspectives.* New York: Seven Bridges, 2000.

Justine Burley and John Harris, eds., *A Companion to Genethics.* Malden, MA: Blackwell, 2002.

Ronald Cole-Turner, ed., *Human Cloning: Religious Responses.* Louisville, KY: Westminster John Knox, 1997.

Sandy Fritz, *Understanding Cloning.* New York: Warner Books, 2002.

Mae-Wan Ho, *Genetic Engineering: Dream or Nightmare?* New York: Continuum, 2000.

Suzanne Holland, Karen Lebacqz, and Laurie Zoloth, eds., *The Human Embryonic Stem Cell Debate: Science, Ethics, and Public Policy.* Cambridge, MA: MIT Press, 2001.

Christopher Howe, *Gene Cloning and Manipulation.* Cambridge, UK: Cambridge University Press, 1995.

David Jefferis, *Cloning: Frontiers of Genetic Engineering.* New York: Crabtree, 1999.

Leon R. Kass, *Life, Liberty, and the Defense of Dignity: The Challenge for Bioethics.* San Francisco: Encounter Books, 2002.

Leon R. Kass and James Q. Wilson, *The Ethics of Human Cloning.* Washington, DC: AEI, 1998.

Arlene J. Klotzko, ed., *The Cloning Sourcebook.* New York: Oxford University Press, 2001.

Gina Kolata, *Clone: The Road to Dolly and the Path Ahead.* New York: William Morrow, 1998.

Paul Lauritzen, *Cloning and the Future of Human Embryo Research.* New York: Oxford University Press, 2002.

Barbara MacKinnon, ed., *Human Cloning.* Chicago: University of Illinois Press, 2000.

Gary E. McCuen, *Cloning: Science and Society.* Hudson, WI: Gary E. McCuen, 1998.

Glenn McGee, ed., *The Human Cloning Debate.* Berkeley, CA: Berkeley Hills Books, 2002.

Robert G. McKinnell, *Cloning: A Biologist Reports.* Minneapolis: University of Minnesota Press, 1979.

Martha C. Nussbaum and Cass R. Sunstein, eds., *Clones and Clones: Facts and Fantasies About Human Cloning.* New York: W.W. Norton, 1998.

Gregory E. Pence, *Flesh of My Flesh: The Ethics of Cloning Humans.* Lanham, MD: Rowman & Littlefield, 1998.

———, *Who's Afraid of Human Cloning?* Lanham, MD: Rowman & Littlefield, 1998.

Lammenranta Rantala and Arthur J. Milgram, eds., *Cloning: For and Against.* Chicago: Open Court, 1998.

Michael Ruse, *Cloning: Responsible Science or Technomadness?* Amherst, NY: Prometheus Books, 2000.

Gregory Stock, *Redesigning Humans: Our Inevitable Genetic Future.* Boston: Houghton Mifflin, 2002.

Brian Tokar, ed., *Redesigning Life?* New York: Zed Books, 2001.

James D. Watson and John Tooze, *The DNA Story: A Documentary History of Gene Cloning.* San Francisco: W.H. Freeman, 1981.

Periodicals

Paul Berg, "Progress with Stem Cells: Stuck or Unstuck?" *Science*, September 14, 2001.

Adriel Bettelheim, "Cloning by Any Other Name: A Defining Battle," *CQ Weekly*, June 15, 2002.

Nell Boyce and James M. Pethokoukis, "Clowns or Cloners?" *U.S. News & World Report*, January 13, 2003.

Mary Agnes Carey and Adriel Bettelheim, "Brownback, Other Ban Supporters Vow to Press for Cloning Debate," *CQ Weekly*, June 15, 2002.

Jon Cohen, "Can Cloning Help Save Beleaguered Species?" *Science*, May 30, 1997.

Jennifer Couzin, "Inheriting Senescence," *U.S. News & World Report*, June 7, 1999.

Economist, "Copy or Counterfeit?" January 4, 2003.

Juliet Eilperin and Rick Weiss, "House Votes to Prohibit All Human Cloning," *Washington Post*, February 28, 2003.

Amy Fagan, "Cloning Issue Again Before Senate," *Washington Times*, January 13, 2003.

Peter Hatfield, "Cloning's Maverick Goes East," *New Scientist*, December 12, 1998.

Leon R. Kass, "The Public's Stake," *Public Interest*, Winter 2003.

Faith Keenan, "Cloning: Huckster or Hero?" *Business Week*, July 1, 2002.

Daniel J. Kevles, "Cloning Can't Be Stopped," *Technology Review*, June 1, 2002.

Jane Lampman, "Cloning's Double Trouble," *Christian Science Monitor*, August 13, 1998.

Mary Ann Liebert, "The Asexual Revolution of Dolly the Lamb," *Journal of Women's Health*, November 1, 1998.

Glenn McGee, "Cloning, Sex, and New Kinds of Families," *Journal of Sex Research*, August 1, 2000.

Celeste McGovern, "Brave New World," *Report Newsmagazine*, June 24, 2002.

Neil Munro, "Doctor Who?" *Washington Monthly*, November 2, 2002.

National Catholic Reporter, "Many Oppose Human Cloning," October 22, 1999.

National Right to Life Committee, "President Bush Urges Senate to Pass the Ban," *National Right to Life News*, March 7, 2003.

Elizabeth Pennisi, "The Lamb That Roared," *Science*, December 19, 1997.

Robert Pollack, "Stem Cells, Therapeutic Cloning, and the Soul," *Cross Currents*, Spring 2002.

Ramesh Ponnuru, "Clone Wars, Part 2," *National Review*, July 1, 2002.

Antonio Regalado, "The Troubled Hunt for the Ultimate Cell," *Technology Review*, July/August 1999.

Sarah Richardson, "Forever Young," *Discover*, January 1999.

Thomas A. Shannon, "Human Cloning: A Success Story or a Tempest in a Petri Dish?" *America*, February 18, 2002.

Peter Singer, "The Year of the Clone?" *Free Inquiry*, Summer 2001.

Davor Solter, "Dolly Is a Clone and No Longer Alone," *Nature*, July 23, 1998.

Bert Vogelstein, Bruce Alberts, and Kenneth Shine, "Please Don't Call It Cloning!" *Science*, February 15, 2002.

Elizabeth Weise, "Road Map Is Drawn for Cloning Primates," *USA Today*, April 13, 2003.

Ed Welles, "Who Is Doctor West and Why Has He Got George Bush So Ticked Off?" *Fortune*, April 29, 2002.

Sylvia Pagán Westphal, "Take a Thousand Eggs," *New Scientist*, February 2, 2002.

Ian Wilmut, "Cloning for Medicine," *Scientific American*, December 1998.

Web Sites

BioSpace, www.biospace.com. BioSpace has "breaking news" on cloning with multiple links to articles and Web sites. The site offers press releases and suggestions for books on cloning.

Electronic Journal on Human Cloning, www.humancloning. mccal.org. This site is supported by Genesis Biotech and offers medical information on cloning as well as editorials, discussions, resources, and the latest articles on cloning.

Human Cloning Foundation (HCF), www.humancloning.org. The HCF is a nonprofit site that promotes human cloning technology and other forms of biotechnology. It has essays and news on human cloning and a free newsletter, the *Cloner*, which covers the latest cloning resources and articles.

National Bioethics Advisory Commission (NBAC), www.bioethics. gov. The NBAC is a federal agency created by President Bill Clinton to research the issues of human cloning. The site offers transcripts of NBAC debates as well as recent news about cloning and other bioethical issues. It has published reports, including *Cloning Human Beings* and *Ethical Issues in Human Stem Cell Research.*

New Scientist.com, www.newscientist.com. This site presents information on the latest advances in cloning technology in agriculture, medicine, psychology, and law. The site has a frequently asked questions (FAQ) section and media information on cloning.

Reasononline, http://reason.com. Reasononline offers numerous articles on the question of whether cloning should be banned. It gives a time line of cloning advances from 1000 B.C. to the present.

ReligiousTolerance.org, http://religioustolerance.org. This site discusses the various types of human cloning, how therapeutic cloning is done, and the ethics, public opinion, and legislation involved in cloning. It also offers the most recent cloning developments.

Reproductive Cloning Network, www.reproductivecloning.net. This is a neutral site offering cloning information. It hosts over sixty cloning Web sites and publishes the latest reproductive cloning resources from many respected authors.

INDEX